BOOMTOWN SALOONS

Wilbur S. Shepperson Series
IN NEVADA HISTORY

BOOMTOWN SALOONS

ARCHAEOLOGY AND HISTORY IN VIRGINIA CITY

KELLY J. DIXON

UNIVERSITY OF NEVADA PRESS
Reno and Las Vegas

Wilbur S. Shepperson Series in Nevada History
Series Editor: Michael Green
University of Nevada Press, Reno, Nevada 89557 USA

Manufactured in the United States of America
Design by Kathleen Szawiola
Library of Congress Cataloging-in-Publication Data
Dixon, Kelly J., 1970–
Boomtown saloons : archaeology and history in Virginia City /
Kelly J. Dixon.
p. cm. — (Wilbur S. Shepperson series in Nevada history)
Based on author's thesis (Ph. D.)—University of Nevada.
Includes bibliographical references and index.
ISBN 0-87417-608-5 (hardcover : alk. paper)
1. Virginia City (Nev.)—Antiquities. 2. Archaeology and history—Nevada—
Virginia City. 3. Frontier and pioneer life—Nevada—Virginia City. 4. Bars
(Drinking establishments)—Nevada—Virginia City—History—19th century.
5. Historic sites—Nevada—Virginia City. 6. Material culture—Nevada—Virginia City.
7. Excavations (Archaeology)—Nevada—Virginia City. 8. Virginia City (Nev.)—
Social life and customs—19th century. I. Title. II. Series.
F849.V8D595 2005
647.95793'56'09034—dc22 2004023025
The paper used in this book meets the requirements of
American National Standard for Information Sciences—
Permanence of Paper for Printed Library Materials, ANSI
Z.48–1984. Binding materials were selected for
strength and durability.

15 14 13 12 11 10
5 4 3 2
ISBN-13: 978-0-87417-703-9

This book is dedicated to

RON JAMES *for having the vision,*

to DON HARDESTY *for paving the way,*

and to CARRIE SMITH *for knowing what*

I needed to make it happen.

CONTENTS

LIST OF ILLUSTRATIONS

Graphs

Maps

PREFACE

THIS IS A BOOK about saloons. More specifically, it is an outgrowth of my doctoral dissertation on the archaeology of a nineteenth-century African American business known as the Boston Saloon, in Virginia City, Nevada. Initially, I planned to turn that research into a book that underscored this enterprise alone.

To place that study in a context of other contemporaneous drinking houses in Virginia City, I took advantage of the database of archaeological investigations in Virginia City. Between 1993 and 1995, Donald L. Hardesty, of the University of Nevada, teamed up with the Nevada State Historic Preservation Office (SHPO) to excavate archaeological remains of two Irish-owned saloons in a disreputable part of town. By the late 1990s I began working with the Nevada SHPO, the Comstock Archaeology Center, and a crew of extraordinary volunteers to investigate the ruins of a fancy German-owned opera house saloon. After adding the Boston Saloon to this mix while working on my doctorate at the University of Nevada, Reno, I found myself comparing a handful of nineteenth-century saloons that highlighted a range of ethnic and socioeconomic affiliations. This story had the potential to accentuate a shared heritage in the West, which could, in turn, cultivate respect for our diverse past. It became clear that the real narrative of boomtown saloons could not come from the view of a single drinking house. I needed to tell the story of not just one, but several saloons.

This mission proved to have the power to unravel enduring glamorized imagery associated with these icons of Western history, imagery that had been perpetuated by Hollywood portrayals and by sensationalism. The physical traces of this group of drinking places symbolized authentic snapshots of life in a dynamic urban setting in the mining West. Indeed, the mining boomtown experience meant and became different things for different people. If multiple stories, such as those of various saloons, can be recovered and told, then a more complex and vivid Western story will emerge.

Virginia City is a living ghost town, with a few strips of buildings displaying faded Victorian-era splendor and clinging to the rugged backdrop of Mount Davidson. Mine dumps create artificial hills amid the desert scrub-covered mountains, marking spots where someone burrowed into the mountainside in search of gold and silver. Attracted to the region's mining wealth, people came here from all over the world during the latter nineteenth century. Virginia City therefore sported a cosmopolitan, diverse population with more than 20,000 people of various cultural and ethnic backgrounds coming into contact with each other in this community on the slope of Mount Davidson.

One of the major tenets of historical archaeology seeks to rediscover and/or reconstruct the histories of the underdocumented lives spent in such diverse communities. Aware of this, I drove into Virginia City in 1997 realizing that historical archaeology had a crucial role to play in understanding the complex story of this boomtown's mining heyday. I did not yet, however, realize that Virginia City's buried resources held such precious untold stories about Western history. Nor did I realize that those stories would emerge from places as notorious as saloons.

ACKNOWLEDGMENTS

THE DAY I delivered this book's manuscript to the University of Nevada Press was the same day I drove away from northern Nevada to start a new job at the University of Montana. In doing so, I left behind a place that had felt like home for many years, and I also bade farewell to numerous people who, in their own ways, moved me to write this book.

The collective influences of Ron James and Don Hardesty head the list. I first saw them in a ballroom at the grand gold rush–era National Hotel in Nevada City, California, at a mining history conference. Just over two years later, Ron James became my supervisor when I worked for the Comstock Historic District Commission, a satellite agency of the Nevada State Historic Preservation Office. Working with Don Hardesty, we soon formed the Comstock Archaeology Center. After crossing paths with Don Hardesty through this organization, I became inspired to earn my Ph.D. with him and asked if he would advise me as a doctoral student. Hardesty and James then became part of a memorable dissertation committee, which kept me sharp while I worked on the research that led to this book. Other members of that committee are Scott Casper, Catherine Fowler, Don Fowler, and Gene Hattori.

Staff members at the University of Nevada Press taught me more than a few details about the publication process while making the work seem effortless. I am especially grateful to Margaret F. Dalrymple, Joanne O'Hare, Sara Vélez Mallea, and Carrie House. Ron James gave unconditional inspiration and editorial advice on every facet of this book, and Gene Hattori came to the rescue to review a newly added chapter when I was at the point of deadline on the manuscript. Jan McInroy is responsible for the final copyedits and polish that transformed the manuscript into a book. To the manuscript's anonymous reviewers, I am most grateful for corrections and suggestions, and I attempted to address them all.

Archaeological endeavors on the Comstock have been supported by a series of generous grants from the Nevada State Historic Preservation Office. The National Endowment for the Humanities provided financial backing for research associated with those endeavors, funding the visits of scholars such as Paul Mullins from Indiana University and Purdue University, and Adrian and Mary Praetzellis from Sonoma State University. Thanks also to Karen Wells.

For their camaraderie and vision, I am grateful to the staff of the Nevada State Historic Preservation Office: Alice Baldrica, Mella Harmon, Ron James, Susie Kastens, Terri McBride, Rebecca Ossa, Rebecca Palmer, and Barb Prudic. Michael "Bert" Bedeau, Deborah Thomas, and Candace Wheeler of the Comstock Historic District Commission relayed telephone messages and helped out with administrative facilities "on the hill" for the duration of my research on saloons. Additionally, Ron James spent countless hours of his own time preparing numerous artifact photos for this book, which, thanks to his photographic skills, is a better product.

The McBride family temporarily gave up their parking lot at the Bucket of Blood Saloon and patiently put up with a constant presence of dust-covered people on their property. Special thanks to Don McBride Sr. and Marshall McBride for their extra assistance with the logistics of excavating the Boston Saloon. Also, cheers to the Bucket of Blood staff.

Marilou Walling, the Storey County Commissioners, Storey County Public Works, and the Storey County Sheriff's Department helped with grant administration, asphalt removal, and security issues. Additional Virginia City support came in various forms from Joe and Ellie Curtis of Mark Twain Books, Jim Reed and the Nevin House, Andria Daley-Taylor, Bo Johanson,

Piper's Opera House Programs, the Delta Saloon and Sawdust Corner Restaurant, Julie Lee and the Virginia City Chamber of Commerce, and John McCarthy.

The project's technical advisory committee provided guidance with planning and public outreach: Michael S. Coray, special assistant to the president for diversity at the University of Nevada, Reno; Lucy Bouldin, director of the Storey County Library, Virginia City; Ken Dalton of the Reno-Sparks NAACP; Elmer Rusco, professor of political science, University of Nevada, Reno; and Theresa Singleton, Syracuse University, Syracuse, New York. Thanks also to the Comstock Archaeology Center's Technical Advisory Board: Ken Fliess, Don Hardesty, Gene Hattori, Ron James, David Landon, Pat Martin, Susan Martin, and Ron Reno.

Comstock Archaeology Center and University of Nevada, Reno (UNR) field crew members gave much energy to the project, and each provided unique expertise: Morgan Blanchard, Larry Buhr, Gary Estis, Patricia Hunt-Jones, William "Jerry" Jerrems, Robert Leavitt, Signa Pendegraft, and Lisa Rizzoli. Comstock Archaeology Center lab crew members Jessica Escobar, Elyse Jolly, and Lorraine Plympton simultaneously implemented a complex database to assist in the artifact analyses for this book, and more importantly, they gave me priceless friendship.

Comstock Archaeology Center education director Dan Kastens lifted public archaeology to higher levels by taking on a challenge the rest of us feared—organizing the visits of more than three hundred children to the field and lab areas and conducting a series of lyceums at various schools. Thanks to the numerous children and school groups who graced the project with their brief but helpful visits. And special thanks to one particularly remarkable young student, Kinsey Kruse, and to her mother, Chris Kruse.

In addition to the children, this project's volunteer crew included several young adults and adults. By taking time out of their work week, vacations, and retirement, these people created the project's enthusiastic workforce and reminded me how to enjoy archaeology. I extend my gratitude to

Barbara Arnold, Alex Atreides, Karen Birk, Lee Brockmeier, Elizabeth Bugg, Jennifer Callahan, Ellie Curtis, Joe Curtis, Kenny Dalton, Susan Damask, Connie Davis, Jeri DeJonge, Ruby DeVos, Alexis Dillon, Cal "my-favorite-guy-on-the-Comstock" Dillon, Erin Elsinger, Al Ferrand, Oyvind Frock, Katie Harris, Heidi Hoff, Karen Hopple, Ken Hopple, Don Ivey, Reed James, Susan James, Matt Jared, Suzy Johanson, Elyse Jolly, Jake Kenneston, Blythe Kladney, Martha Kraemer, Kaley Kruse, Kelly Lang, Nancee Langley, Janice Leavitt, Lynn Leavitt, Catherine Litz, Christian Malone, Jennifer Manha, Terri McBride, Tim McCarthy, Sherry McGee, Valorie Morgan, Christopher Nelson, Matt Potts, Barb Prudic, Shawn Rowles, Elmer Rusco, Sheryl Sanders, Ed Smith, Emily Sparks, Robin Sparks, Shane Sparks, Lance Taylor-Warren, Lori Taylor-Warren, Zack Taylor-Warren, Alexandra Toll, Nicole Tyler, Ryan Tyler, Michael Weidemann, Alex White, Chris White, Robert Wolfe, Josephine Wong, and Andy Zogg.

Two people deserve extraordinary recognition as volunteers. Dan Urriola volunteered 2,100 hours sorting and mending hundreds of shattered ceramics and glass from the Boston Saloon. His dedication to this project inspired me to be positive during moments of frustration. He is the lifeblood of the project and the reason the lab work was completed so quickly. Christine Urriola also warrants admiration for the long-term loan of her husband to Virginia City archaeological projects, for tolerating the transformation of her home into a ceramic-mending facility, for taking numerous artifact photos, and for figuring out how to feed me.

The students enrolled in the UNR archaeological field school at the Boston Saloon worked as competently as professionals to recover and record saloon artifacts mentioned in this book: Brian Alcorn, Len Balutis, Shari Davis, Amber Devos, Gene Dimitri, Leilani Espinda, Trish Fernandez, Troy Garlock, Chris Knutson, Robert Leavitt, Tina Pitsenberger, Lorraine Plympton, Kelly Seaton, Kathy Sharkey, Maie Tsukuda, and Diane Willis. Thanks to Ahern Rentals and to Sierra Springs for making the field school students' jobs a little easier with heavy equipment and drinking water donations.

Other people provided comments, ideas, inspirations, and general academic support while I worked on the research that led to this book; their assistance and encouragement are not forgotten: Scott Baxter, Morgan Blanchard, Cris Borgnine, Marie Boutte, Alyce Branigan, Larry Buhr, James Davidson, Barbara Erickson, Ken Fliess, Darla Garey-Sage, Ann Harvey, Dave Harvey, Greg Haynes, Erika Johnson, Dean P. McGovern, Robert Leavitt, Tim A. Kartdatzke, Robert Kopperl, Robert W. McQueen, Margo Memmott, A. Millard, Paul Mullins, Adrian Praetzellis, Mary Praetzellis, Elmer Rusco, Jessica Smith, Cathy Spude, Laurie Walsh, and Robert Winzeler. I am especially grateful to Stephanie Livingston, a professional who volunteered so much of her time to assist with faunal analysis on the Boston Saloon and Piper's Old Corner Bar. G. Richard Scott inspired me to keep pressing on, and he continues to do so.

Archaeologists and employees from numerous government agencies assisted on many levels during my time digging saloons in Virginia City. Maggie Brown, Gene Hattori, Larry Tanner, Alanah Woody, and Roz Works of the Nevada State Museum provided exhibit ideas, temporary storage, and access to faunal remains. Special thanks to Jan Loverin of the Marjorie Kemp Textile Museum, Carson City, Nevada.

Saloon archaeology began in Virginia City several years before I arrived there. That means there are countless people whom I have not met, including University of Nevada, Reno field school students, who participated in some projects mentioned in this book. I extend thanks to them for recovering portions of the saloon history.

Julie Schablitsky provided references, brainstorms, and friendship throughout the duration of this research; additionally, the forensic chapter is the direct result of her influence and ideas. Stacy Schneyder-Case has always been inspiring, and I can only hope that some portions of this book exude her excitement for the discipline of archaeology and archaeologists.

Richard Paul "Benny" Benjamin provided crucial comments on various drafts and reminded me how to live. While I was writing this book on a "mini-sabbatical" in Amsterdam, an unforgettable circle of experts demonstrated the finer points of leisure, a topic with which I have long struggled: Richard P. Benjamin, Elizabeth Anne Knight, Malcolm Plaister, Vicki Rudd, Maud van Waardhuizen, Jason Walsh, Simon Whitehall, and Mat Wilson.

After bidding farewell to a second family, friends, and colleagues in northern Nevada, I soon found a new home at the University of Montana. My colleagues at the Department of Anthropology there impressed me with a professional network of support that gently eased me into my first year as a new professor and that allowed me time to polish the manuscript for this book. Big thanks to a stellar department: Greg Campbell, John Douglas, Tom Foor, Stephen Greymorning, Kimber Haddix-McKay, Gary Kerr, Linda McLean, Bill Prentiss, Dick Sattler, Noriko Seguchi, Randy Skelton, and G. G. Weix. I also send gratitude to Gerald Fetz, Dusten Hollist, Bill McBroom, Becky Richards, and Celia Winkler.

I thank the students enrolled in my fall 2003 historical archaeology course at the University of Montana for their input, inquisitiveness, and enthusiasm: Kristin Bowen, Jen Childress, Sarah Cole, Daniel Comer, Brooke Easterday, Jonathan Hardes, Danielle Marcetti, Dirk Markle, Julie Marshall, Jack McShane, Chris Merritt, Lisa Weeks, and Ben Woody.

Carrie Smith and Paul Spencer gave me food, a bed, and a writing refuge on countless occasions. They also made sure I had the opportunity to escape all other responsibilities for a three-month period, thus ensuring the preparation of the initial manuscript for this book.

For warm meals and a home while I scurried from university commitments to boomtown diggings, I am ever appreciative of my second family: Charles, Diana Lynn, and Justice Malone, and July Wright. Christian Malone deserves special recognition for his behind-the-scenes care and support with all of my work in Virginia City.

For their unconditional encouragement and patient understanding of my passion for the past, thank you with love to my family: Laverne Bonstrom, Tracy and Bob Davies, Cami Dixon, J. David and Jeanne Dixon,

Wendy Dixon-Etzel, Doug Etzel, Elijah Etzel, Neil Shook. Special thanks to my uncle, Jerry Bonstrom, for comments on an early draft of this book's introduction. Finally, Giles C. Thelen provided editorial advice along with an infrastructure for my life, all of which aided the finishing touches of this book.

WITHOUT DOUBT, the information compiled here resulted from the dedication of many people. I take full responsibility, however, for any oversights in the research. While I have attempted to remember as many of those individuals as possible, I realize that it is impossible to account for everyone who made a contribution. To anyone whose name is not mentioned here, I send apologies, with gratitude, and consider myself blessed that there are so many to thank.

BOOMTOWN SALOONS

INTRODUCTION: HISTORICAL ARCHAEOLOGY METHODS

Much More Than Digging with Small Tools

The very notion of archaeology evokes images of a field excavation, with people in brimmed hats bent over a gulf of contiguous pits. On hands and knees those dusty individuals delicately wield tools that are ridiculously out of proportion to the amounts of earth they excavate. Their mission: to discover long-lost and unexpected antiquities. For many, these scenes are fascinating to watch—either in person or on any number of entertainingly educational television programs.

In the face of such allure it is common to hear statements like, "I don't know if I could have the patience to sit in a hole all day and do that." For a unique handful of people, being part of such sincere, meticulous explorations is spellbinding. Much to the astonishment of friends and family, those few decide to make a life of archaeology. To do so, they need to learn a comprehensive range of archaeological methods and be aware of the theory lurking in the background.

Ironically, despite the most careful artifact removal and meticulous excavation, the very act of extracting historic or prehistoric objects from buried (or surface) contexts destroys the essence of an archaeological site.[1] Consequently, people working on archaeological projects have a responsibility to ensure that the nonrenewable settings that contained the remains they unearth are recorded in as thorough a manner as possible.

Archaeologists around the world deal with the frustrating reality that their science is inherently destructive. The trade-offs are nevertheless worth it and culminate in the discovery of long-lost and astonishing details about the human time line.[2] Given this rewarding undertaking, archaeologists continue to seek knowledge about the past and keep checks on the destructive nature of their science. Fieldwork is much, much more than digging. As a matter of fact, the most important qualification that one can bring to an excavation is good recording skills. Photographs, maps, detailed notes, and records chronicling the proveniences of a site's artifacts and features are essential for ethical archaeological fieldwork. The ability to keep impeccable records is a cornerstone of field methods and is an important part of the skills commonly taught to university-level students in archaeological field schools.

Public Archaeology

The concept of public archaeology was initially associated with projects undertaken as part of government-mandated salvage operations. Since the 1980s, members of the general public, through the mission of "public archaeology" and related volunteer programs, have also had the opportunity to learn archaeological field methods. The term came to refer to research that engages the public through volunteer programs, site tours, and cooperative work with descendant communities. Such activities were first developed at historic site museums, such as colonial Williamsburg, Monticello, and Mount Vernon, spearheading the new version of public archaeological research projects.[3] In 1993 public archaeology officially appeared in Virginia City, Nevada.

Since Virginia City and colonial sites date from the historic period, the type of archaeology to which many members of the public have been introduced is one that recovers recognizable objects, such as tobacco pipes. The thrill of being able to identify (and identify with) such artifacts is often juxtaposed against the surprise among the public that people carry out archaeological excavations from our relatively recent past. Archaeology has long been associated with far-off places and with sites of longer antiquity than the historic period in North America.[4] Nevertheless, many of the archaeological methods used at various sites and shared with the public can be applied to a range of ruins, be they ancient and exotic or recent and familiar. Archaeology in Virginia City yielded materials that helped to change the familiar and stereotypical conceptions of the "Wild West."[5]

ONE OF THE MOST STRIKING PLACES in the world, Virginia City, Nevada, is both a National Historic Landmark and a place where people live. Consequently, it has many layers. For example, its exterior cover has the enchanting appearance of a nineteenth-century western boomtown laden with saloons, boardwalks, miner-cowboy look-alikes, and a donkey. The place also has an inner layer of being "home" for many people, and it was home to many of their families before them.

Virginia City therefore became an exceptional place to carry out a public archaeology project that involved thousands of passerby tourists, neighboring residences and businesses, and a local community-based volunteer archaeology fieldwork program. Between 1993 and 1995 archaeologist Don Hardesty teamed up with historian and Nevada State Historic Preservation officer Ron James to develop a public archaeology program in the Virginia City National Historic Landmark. The fact that they initiated this project in the middle of the tourist season—and that the program also served as a training ground for archaeology students—was groundbreaking and created a standard for carrying out archaeological research in that community.

I was fortunate enough to begin research there after this paradigm for public archaeology became established. Two saloons were among the sites excavated by Don Hardesty's crews, and I had the opportunity to excavate two other saloons. Because the previous archaeological research had already blazed a path, I came upon a setting with a precedent for public historical archaeology in the mining West, and it had an existing database of saloon artifacts. This book is merely a part of ongoing archaeological research that is being synthesized with western history, and that subsequently contributes to a revisionist western history.[6] And since this book is about archaeology, it only makes sense to rewind to the part of this story that begins with the quintessential picture of archaeology: the fieldwork.

A series of existing documents, including excavation reports and a doctoral dissertation, already provides descriptions of the fieldwork for each saloon project.[7] Rather than repeat and describe the field methods for each saloon excavation here, which could become tedious, I focus in this section on the methods used at the Boston Saloon, which involved a significant public archaeology charter and represents the most recent of the saloon archaeological projects in Virginia City.

Public Archaeology and Field Methods at the Boston Saloon

The Boston Saloon project was the first excavation of an African American saloon in the mining West. In part, it served as a field school for university-level students, working under the instruction of Don Hardesty, to gain hands-on experience with excavation methods.[8] Additionally, the project emphasized the value of archaeological methods for African American history and for western history. It did not limit this approach to field school students, however, but developed a multifaceted public archaeology program, with more than sixty-five volunteers participating in the fieldwork during the project's five-week period. More than three hundred children came up for brief but helpful visits throughout the summer as well.[9] To efficiently balance this massive public interest, the project's archaeologist and education director had to carefully schedule the visits of groups and volunteers for the summer field season before excavation. This strategy ensured that there were not too many people in too small a dig area.

Another facet of the Boston Saloon public outreach mission focused on the fact that archaeology gives many people—not just those directly involved with a dig—access to tangible remains of the past. The Boston Saloon student and volunteer crew routinely showed and described their findings to thousands of visitors who passed their excavation areas during the course of the five-week excavation. To cater to the people who were curious but who did not approach the excavations too closely, a large sign explaining the project stood at the edge of the site, with brochures about the Boston Saloon project and the role of African Americans in Virginia City attached to it.[10]

Before discussing the public's involvement during fieldwork, it is necessary to back up to the project's planning phase and discuss the public archaeology component at that point. First it was necessary to obtain permission from the owners of private property to conduct a public archaeology project on their land. The McBride family of Virginia City owns the Bucket of Blood Saloon, the property that contained the ruins of the Boston Saloon beneath its asphalt parking lot, and they kindly agreed to let their property be transformed from a parking lot to an archaeological dig during the busy summer tourist season. They agreed—with one major and quite fair condition: that somebody take responsibility for backfilling and repaving their parking lot at the end of the excavation.

After securing permission and appropriate permits, archaeologists do not just go out and randomly plunk holes into the ground in the areas where they believe sites are buried. Rather, they go into the field with a carefully written plan, a "research design." The research design for the Boston Saloon described what was known about that establishment by the spring of 2000, using information that had been gathered primarily from historical records. The design posed a series of questions that could potentially be answered by historical archaeology and then described how the archaeology would be carried out—and by whom—over a certain time line. A public archaeology mission was outlined in the research design, involving an outreach program that would invite African Americans from the area surrounding Virginia City (e.g., Reno, Sparks, and Carson City) to participate in the recovery of their heritage in the mining West, specifically in northern Nevada. The Reno-Sparks chapter of the NAACP aided this effort by distributing flyers about the project at local schools and churches, and the Northern Nevada Black Cultural Awareness Society (NNBCAS) received information packets about the project for their planning and informational purposes. Finally, Ceola Davis, the editor of the monthly newsletter *En Soul,* ran an article about the project in her publication, and this became another crucial level of outreach for the Boston Saloon story.

Continuing the project's charter of public involvement, a number of people from various public and professional backgrounds, including African Americans specializing in history and archaeology,[11] were asked to review and comment on the research design. The intent of this inquiry was to integrate non-Eurocentric ideas into the design, as well as to ensure the project's relevance for the region's African Americans.

After reading a draft of the research design, these individuals made recommendations for how research should be conducted, and their ideas were incorporated into the final plan for the Boston Saloon project.

For example, one of the committee members recommended that security measures be implemented to protect the site before and during fieldwork. Because of a long history of looting in Virginia City, including an incident that took place only a short while before the Boston Saloon was excavated, we were justifiably concerned about uncovering a site that had been capped and therefore well protected from looting by an asphalt parking lot *and* that represented the first of its kind to be examined archaeologically. Thus we explored several options for ensuring the site's protection once the asphalt was removed.

While committee members understood the grave potential for looters to damage freshly uncovered archaeological deposits, one individual expressed an additional concern, explaining that such security measures were essential to protect a heritage that others might seek to damage or erase in the name of racism. She was aware that the project would help to create a new memory in the region that was, in most mainstream circles, not part of the typical western story. This memory of a shared and diverse past in the West was, in the eyes of certain people, better left buried. Perhaps Eurocentrically, I had not even anticipated the possibility of such bigoted, negative reactions to the Boston Saloon project.

Security measures, although costly, therefore became an essential component of the project's public archaeology field methods. The same day that the backhoe lifted the asphalt layer from the Bucket of Blood Saloon's back parking lot, a security fence arrived. The fence surrounded the perimeter of the parking lot until the end of the excavation. The Bucket of Blood Saloon also agreed to have a temporary surveillance camera installed outside one of its upper-story windows. The camera provided twenty-four-hour coverage that, along with the fence, was intended to deter potential looters or vandals. Of course the camera also provided a means of recording and therefore aiding the prosecution of any actual looters and/or vandals.

Historical Research Methods for the Boston Saloon

While the research design outlined the planning process for the Boston Saloon project before anyone put a digging tool into the ground, it would have been impossible to find the right spot to begin excavation without the aid of historical research. Archaeologists researching the historic period would be lost without original records. Although it is necessary to be savvy about the inherent biases of the people who created those accounts, the fact remains that contemporary newspaper articles, maps, photographs, diaries, business directories, tax and property records, and census records are among the major sources that guide archaeological research of historic-period sites.

Such records are the domain of historians, which means archaeologists should consult with those specialists when considering fieldwork on historic sites.[12] Some historians are also aware of how the artifact record can enhance interpretations of the past. For example, soon after the 1997 founding of the Comstock Archaeology Center, Ron James, a historian and a cofounder of that organization, asked for a list of possible historical research topics that would benefit from archaeological research. An investigation of African

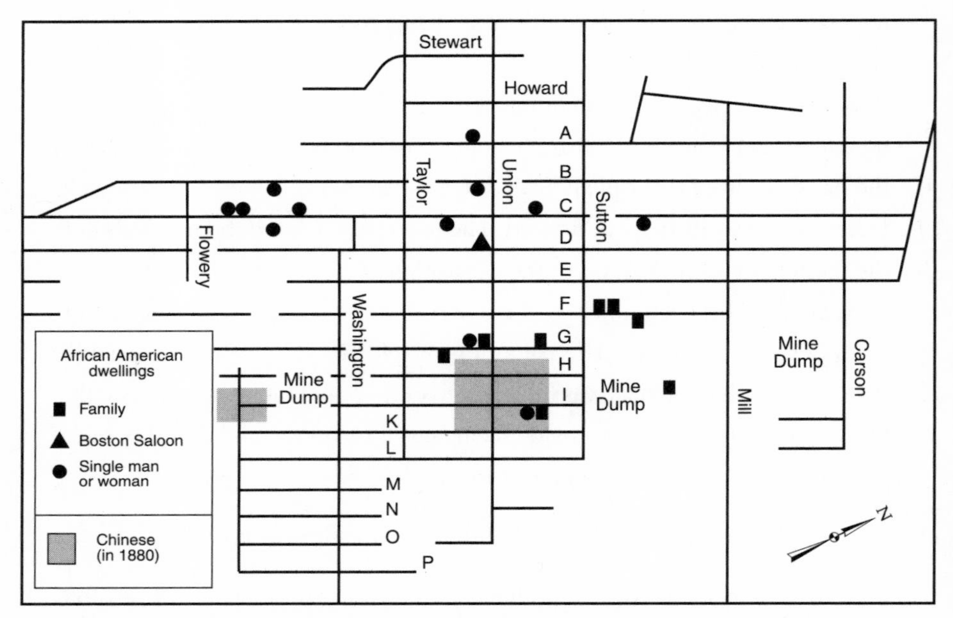

MAP 0.1. Map compiled by Ronald M. James using information from directories, census records, and fire insurance maps to show the various and integrated locations of African American households in Virginia City during 1873–1874. Modified from Ronald M. James, *The Roar and the Silence* (Reno and Las Vegas: University of Nevada Press, 1998), 99; courtesy of the University of Nevada Press.

Americans in the mining West was one of the suggested topics. The scarcity of archaeological investigations of African Americans in that region influenced the decision to prioritize that topic in Virginia City.[13] The initial research goals included a general focus on locating places where African Americans had worked and lived on the Comstock Mining District during the nineteenth century; at this point the research potential of the Boston Saloon was not fully realized.

Many of the kinds of historical records noted above, such as census manuscripts, nineteenth-century business directories, and county records, helped guide the research to find out how many people of African descent had lived in the district during the mining boom and to determine where they lived. We soon learned that African Americans in Virginia City did not settle in a distinct or designated community as did people of Asian ancestry; the latter, most of whom came from mainland China, resided in the neighborhood known as Chinatown. While single African Americans were primarily dispersed along Virginia City's commercial corridor at that time, families lived in scattered locations downhill from that corridor, in areas with low real estate values on the edge of Chinatown and the mine dumps of the Consolidated Virginia Mine, but integrated within Virginia City's diverse community (map 0.1). For example, they shared boardinghouses and neighborhoods with European immigrants and European Americans. Although it is inappropriate to overemphasize the degree of integration, especially given the complexity of ethnic relations that likely existed in Virginia City, it is still noteworthy that former slaves rented rooms from and lived next door to Europeans and European Americans.[14]

While the revelation of integration illustrates at least some sophistication on behalf of Comstock society during the Civil War and the Reconstruction era, the potential for mixed deposits deflated hope for archaeological remains that could accurately be linked with African Americans.[15] Furthermore, the many black-owned business enterprises left few traces of their presence from an archaeological point of view because they frequently changed locations. Therefore, the historical records initially suggested that it would be a difficult task to identify African Americans in Virginia City's archaeological record.

After correlating several historical references, Ron James homed in on the site of the Boston Saloon. Multiple lines of evidence, including newspaper articles from the *Territorial Enterprise*, the *Virginia and Truckee Railroad Directory* of 1873–1874, Nevada State Census records from 1875, and Sanborn-Perris Fire Insurance maps all pointed to the location of a saloon that served as the "popular resort for many of the colored population" and that was owned by African American William A. G. Brown.[16]

FIG. 0.1. Parking lot on the east side of the Bucket of Blood, capping the Boston Saloon. Photo by Ronald M. James

Oral histories are another valuable historical resource that can provide a layer of meaningful depth for archaeological excavations. Although such material could potentially have given voice to African Americans in northern Nevada's late-nineteenth- or early-twentieth-century mining communities, the records were lacking.[17]

After the initial research had exhausted the possibility for oral histories and a nationwide press release requesting information about African Americans in Virginia City, Nevada, had been issued, it was clear that only a few lines of historical evidence were available to guide archaeologists in their interpretations of an African American saloon. These records indicated that the Boston Saloon operated at Number 4 South D Street, the southwest corner of D and Union Streets in Virginia City. The long-lived Boston Saloon stayed at that single location throughout most of its existence. This stability made it an anomaly among small businesses in a mining boomtown and—more important for archaeologists—created a site with sufficient longevity to warrant archaeological investigation.

The current location of Number 4 South D Street is the asphalt parking lot, spanning 18.5 meters by 15.5 meters (61 feet by 51 feet), that sits on the east side of the Bucket of Blood Saloon. In addition to the Boston Saloon, the parking lot covered at least four other structures that were situated on the southwestern corner of D and Union Streets during the period between the 1860s and the 1870s. The site was a garden during the 1930s, a dirt parking lot from the 1950s to the late 1970s, and an asphalt parking lot since about 1980 (figure 0.1).[18]

Field Methods at the Boston Saloon

Asphalt lots present pretty significant barriers for people who usually dig with trowels. Even so, this difficulty was a minor obstacle compared with the profound potential of recovering the tangible heritage of a group of people who had been excluded—for the most part—from the archaeological stories of the Wild West. We had only to figure out how to remove and replace the parking lot in the name of archaeology. The solution to the problem required some relatively unconventional archaeological techniques, such as using heavy equipment.

Before storming the site with heavy equipment, archaeologists needed to dig a test pit. This crucial prelude to a major excavation helps to determine the nature of buried deposits and provides a preview of the stratigraphic, or layered, appearance of a site.[19] Fortunately, a small patch of ground in the parking lot had not been paved over, so this area was selected as the site for a test pit in 1998, two years before the official parking lot removal and excavation.[20] The test pit revealed a colorful contrast of layers in the earth

FIG. 0.2. Rich Backus of Storey County Public Works maneuvering the backhoe to remove asphalt from Bucket of Blood parking lot.

beneath the parking lot, including a band of gray ash. Beneath that lay a distinct blackened layer, full of charred wood fragments and burned, disfigured splinters from broken glass objects. Just beneath the burn layer, a tiny fragment from the pedestal of a crystal goblet emerged. When combined with Virginia City history, these layers revealed a story that was about to make the pending excavation of the Boston Saloon even more thrilling. The burn layer was a blatant reminder of Virginia City's terrifying and massive Great Fire of 1875.

In the case of the Boston Saloon, that grayish-black temporal marker took on deeper meaning because the establishment's proprietor, William A. G. Brown, had closed his saloon in 1875, just months before the Great Fire, and the fire layer literally capped the remains of the Boston Saloon. Because Brown's establishment had operated for nine years at that single location, material traces of the saloon, such as the pedestal from a crystal wineglass, were lost, thrown out, and subsequently built up in tiny layers until they were covered by the charred wood and ash of the 1875 fire. Realizing this, archaeologists knew that the remains of the saloon were protected and likely in pristine condition beneath the parking lot. The results of the test excavation deemed it worthwhile and quite essential to undertake a major excavation, starting with the removal of the entire parking lot.[21]

The first step in the excavation of the Boston Saloon, then, was to use a backhoe to remove the parking lot (figure 0.2). The backhoe carefully peeled back the asphalt without disturbing any of the earth beneath it. Artifacts immediately became visible, and so the archaeology crew took a conservative position and did not use the backhoe to dig any deeper.

After the removal of the asphalt, the crew established a 1 x 1 meter grid for an open-area excavation, set up a screening area, and erected a tent for the field laboratory. The one-meter squares helped to maintain strict control within the open-area excavation (map 0.2). The grid spanned a 6 x 12 meter area on the ground beneath the parking lot that coincided with the site of the Boston Saloon.

The crew did not excavate several portions of the site, leaving a 1 x 6 meter baulk[22] intact in the center of the excavation grid and a larger, 2 x 6 and 3 x 6 meter baulk along the east-central edge of the grid as a means of preserving portions of the Boston Saloon site for future studies. They also avoided the westernmost portion of the site in the name of additional site preservation and to prevent undermining the Bucket of Blood structure; the field lab was subsequently set up in this area. The crew also avoided the southeastern portion of the parking lot because of time constraints and because the area lay outside the Boston Saloon site. Another unexcavated area was a narrow strip outside the saloon's northern border. The crew avoided this area because of time constraints, because it was not thought to be part of the Boston Saloon site, and because the area was used for screening stations.

As the crew members commenced hand excavation, they realized that the artifacts immediately beneath the asphalt represented displaced deposits in

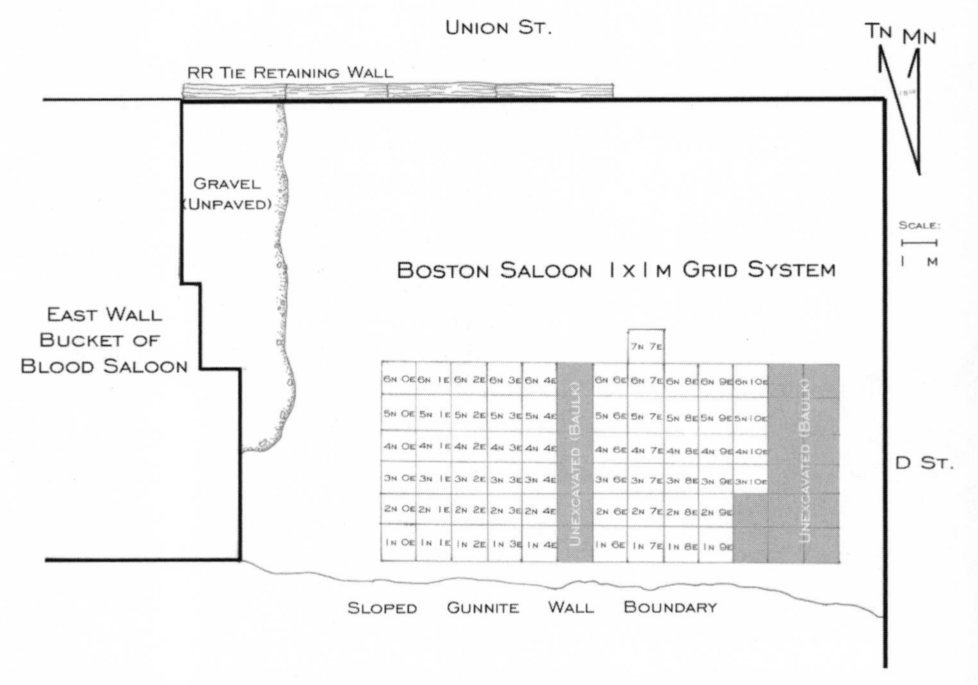

Map 0.2. Site map and layout of grid system at the Boston Saloon after asphalt removal.

a fill layer. The fill was not part of the site's historic deposits and had obviously been brought in to construct the asphalt parking lot. Once the crew realized how deep the fill layer was, they used a "Bobcat" to peel away the fill material in order to expedite excavation (figure 0.3).[23] Once the fill was removed, the crew continued with careful hand excavation in the historic deposits (figure 0.4).

The crew dug in stratigraphic units and pedestaled artifacts in situ within each distinct stratigraphic deposit or surface as a means of assuring that the field school students developed careful excavation techniques. *Pedestaling* refers to leaving artifacts where they were uncovered, atop tiny pedestals of earth (figure 0.5). Crew members then mapped and photographed each stratigraphic unit and its respective artifact deposits before moving on to

Fig. 0.3. Crew member Larry Buhr operating the "Bobcat" in a portion of the fill layer.

Below:

Fig. 0.4. After the heavy equipment had removed the asphalt and the fill layer, the Boston Saloon archaeology crew commenced careful hand excavation. One baulk is shown as an unexcavated strip in the center of the open area excavation near the right side of this photo. Other unexcavated areas are to the left and top of the excavation shown here; note the screening stations and mounds of "backdirt" on the edge of the excavated area at the top of this photo.

underlying contexts (figure 0.6). That way, even though the crew removed major portions of the saloon's nineteenth-century deposits, it would be feasible to reconstruct the site by fitting together all of the crew's notes, maps, and photos. While digital photography was available at the time of the Boston Saloon excavation during the summer of 2000, the crew decided to fall back on the more conservative and traditional use of manual cameras with color slides and black-and-white print film.[24]

Even though the excavation crews carefully mapped and photographed hundreds of artifacts as they found them, small items, such as thin glass shards and tiny beads, were often caught by trowel cuts and scooped up in dustpans with ash and clay. These tiny objects were found when the crew carefully screened small fractions of material from their excavation units through one-eighth-inch mesh. The project tracked all excavated artifacts, whether found in situ or in screens, by recording artifact specimen bag and associated provenience information on a master artifact bag sheet.

In addition to keeping strict records regarding the provenience of recovered artifacts, the Boston Saloon excavation employed the Harris Matrix system to document the site's stratigraphic context.[25] The Harris Matrix provides a system for recording the various layers of an archaeological site. Ideally, a site should be excavated by paying attention to the vertical nature of a particular layer and by following the horizontal extent of that layer. That way, an entire stratigraphic context can be excavated as one episode that has a three-dimensional character, and all artifacts from that episode will be collected as part of a clearly defined deposit.

This method is a challenging undertaking in the learning environment of a field school, though, so the archaeologists laid out a 1 x 1 meter grid to provide student excavators with at least some horizontal area control by assigning everyone respective "squares" in which to excavate. Then they excavated by following the change(s) from one stratigraphic layer, or context, to another. Because of the nature of the open-area excavation, the horizontal contexts were still visible across various contiguous excavation units. For

Top: FIG. 0.5. The archaeology crew left artifacts in place (in situ) while they excavated. After completing the exposure of a stratigraphic layer, the crew mapped and photographed the layer.

Bottom: FIG. 0.6. Field school students Lorraine Plympton and Gene Dimitri map artifacts in one stratigraphic layer before collecting the objects and moving on to deeper excavation.

Fig. 0.7. The 1875 fire lens appears as the gray area above and spreads across various excavation units.

example, the thin 1875 fire layer clearly appeared as a horizontal context; as the crew dug beneath that context, they exposed its vertical context, or depth (figure 0.7).

The Harris Matrix system uses the term *context* as a generic term to describe the various types of stratigraphic units that excavators encounter. For example, a context may represent a *deposit* (e.g., ash or clay) or it may represent something like a surface (e.g., floor or ditch within a building foundation). *Context* also refers to an interface, which is the dividing line between deposits; these dividing lines are sometimes described as the surfaces of each subsequent deposit.[26] This means that a *surface* is often referred to as an *interface* when describing the point of contact between deposits. The methods employed for recording the stratigraphic units of the Boston Saloon site represented an attempt to designate a separate context number for each layer and each surface/interface between each layer to provide a separate record for each discrete unit of stratification (figure 0.8).[27]

The site's stratigraphic makeup started with a line of fill immediately beneath the asphalt. The fill blanketed the topmost part of the site and was between 25 and 30 centimeters, or roughly one foot, thick. Artifacts contained within this fill included nineteenth-century objects and modern materials such as candy wrappers and plastic drinking cups. While the fill represents a modern occurrence that was necessary for paving the parking lot, the presence of historic artifacts is a clue that the fill itself likely came from somewhere in or around Virginia City.

Beneath the fill, the site's layers revealed two general activity areas. One is the westernmost portion of the excavation grid, where there was a dump and an alley behind the saloon structure. In general, artifacts dating from the 1890s and early 1900s appeared in the shallower portion of the dump, giving way to items from the 1860s and 1870s below. The deepest materials in the dump rested on bedrock, which was 60 centimeters below the surface,

FIG. 0.8. Small tags mark each stratigraphic context as observed in the central portion of the Boston Saloon site. Note the gray ashy layer toward the base of this excavated area, which represents the 1875 fire deposits.

demarcating the end of historic cultural deposits in that area. On the basis of the artifact density and dates of the materials, it could be concluded that people threw trash in this dump from the early days of Virginia City's development up to the early twentieth century when the McBride family transformed the area into an orchard. Ash deposits appeared in association with the upper and lower dump deposits and likely represent activities such as burning garbage in the alley. Structure fire probably helped create these two burn deposits, including the well-documented Great Fire of 1875.

A distinct ash deposit appeared in the eastern portion of the site as well. This second portion of the site held the saloon structure, with charred but intact floorboards portraying the most visible remains of that structure. The ash layer and floorboards were 40 centimeters below the surface, beneath a 10-to-15-centimeter-thick deposit of building materials and late-nineteenth-century artifacts. The latter likely represent materials associated with post-1875 occupation of the site, because the ash layer and charred floorboards denote building debris from the 1875 fire.

Because the Boston Saloon closed its doors shortly before that fire, the ashy matrix, as suggested by the pre-excavation test pit, provided a noticeable temporal marker for distinguishing the Boston Saloon from other activities that took place at the corner of D and Union Streets. This is significant because of the complex nature of archaeological deposits in urban settings; businesses move in and out of buildings, and new buildings are constructed atop previous structures in such settings. Indeed, this was the case in Virginia City, which means many activities took place at the corner of D and Union Streets. It was important for the excavation to determine which stratigraphic deposit could be associated with the Boston Saloon's lengthy nine-year tenure at the street corner.

Artifacts from the layer immediately beneath the 1875 fire layer could confidently be associated with the pre-1875 operation of the Boston Saloon at that street corner. Artifact dates correlated with the latter half of the nineteenth century, with a paucity of modern intrusions, providing a confident date range for the stratigraphic layers that held the ruins of the saloon. The fact that the bulk of those artifacts included liquor bottles, glassware, and a handful of food serving vessels provided another type of data, since such items imply saloon activities during the latter portion of the nineteenth century. This evidence, in turn, instilled confidence that the material recovered and analyzed did indeed represent one of the several activities at a bustling Virginia City street corner; and thanks to historical records, it could be verified that the Boston Saloon was one of the operations in that location.

Laboratory Methods and Publication

Contrary to the popular image of archaeologists bent over in dusty pits, most modern-day archaeologists spend more time conducting laboratory analyses, writing reports, and publishing data than they spend in the field. The general rule with archaeological lab methods is that at least three days are needed in the lab for every one day spent in the field. The Boston Saloon project proved that this rule could be extended, for it took students and laboratory employees three years to see the project to completion after returning from the five-week field session.

During this time, the laboratory crew sorted artifacts by material of manufacture and initially classified each object or set of objects according to that material (e.g., glass, ceramic, metal, bone). However, classification systems that account only for materials hinder analyses and are problematic when applied to comparative studies.[28] Because of this, the Boston Saloon's artifact classification system, that is, its catalog of artifacts, was designed to identify the traditional functions of the historic artifacts (e.g., nails and window glass have more explanatory power about a site's makeup when they are categorized in a functional "architectural" category instead of according to the sterile but necessary material functions of metal and glass).[29]

The Boston Saloon's artifact catalog was therefore designed to accommodate such details about the tens of thousands of artifacts recovered from that site. This was the approach taken at the beginning of lab work and continued simultaneously with that work. It meant that at the lab in the University of Nevada's Department of Anthropology a massive artifact catalog and database were created. Lab activities included sorting, cleaning, mending, labeling, and analyzing artifacts, and employing forensic testing and faunal analyses. Finally, the cataloged and cleaned artifacts were prepared for storage and exhibits, with storage box numbers built into the artifact database to allow easy access to each item.

Report preparation and publication is considered the final stage of archaeological research, for this is the medium through which archaeologists compile and interpret their discoveries, or more technically, their data. Don Hardesty, with the assistance of others, prepared a report of his saloon investigations at the Hibernia Brewery and O'Brien and Costello's Saloon and Shooting Gallery.[30] Similarly, I prepared a report that detailed the methods and discoveries at Piper's Old Corner Bar.[31] The data compilation and interpretation of the Boston Saloon site became my doctoral dissertation.[32] In addition to creating reports, archaeologists described research from all of the above investigations at professional conferences and are beginning to publish that research in peer-reviewed journals.[33] This book, along with a proposed exhibit, "Havens in a Heartless World," represents other aspects of the final phase of research, with the goal of making the general results of all four saloon studies accessible to a wide audience.

Even after the preparation of a book like this, archaeological methods are really not exhausted. This publication is merely one person's attempt to make sense of a bulk of archaeological materials with the hope of ever so slightly revising western history. The artifacts are still available for exhibits, for further analysis, for different interpretations, and for other stories about the West. If the archaeologist's methods have been truly successful, artifacts and excavation records will provide material for analysis and comparison for decades or even centuries to come.

OPENING SALOON DOORS

Archaeology Unearths the Real Mother Lode

At the beginning of the 1993 film *Tombstone,* a narrator notes that the murder rate in the sparsely populated "frontier West" was higher than anywhere else in the United States, including New York City. A few minutes later, the scene pans to Tombstone, Arizona, shown as a dusty western boomtown with false-front mercantile buildings. The panorama effectively displays the juxtaposition of finely dressed ladies and gentlemen against the backdrop of grit and desert. A Chinese immigrant hustles among the crowd, a testament to the filmmaker's historical research. In the foreground, Wyatt Earp—played by Kurt Russell—chats with his brothers and a sheriff about saloons. During their conversation, these drinking establishments are described as "the real mother lode" in the growing community. While the filmmakers demonstrate their attention to historical accuracy by including this poignant statement about the significance of saloons, they clearly could not resist the more popular, sensational caricature of these businesses, for the scene closes with the sound of gunfire and a group of grimy men tumbling onto the street from a saloon. *Tombstone* provides an example of the ways in which the western film genre links violence with saloons. Admittedly, films and television programs have created national representations of "the West," with popular, sensationalized depictions elevating the saloon to its status as a place of hostility and as a conveyor of Americana.

The popularity of television shows such as *Gunsmoke* and *Bonanza*

FIG. 1.1. J. Ross Browne's "Home for the Boys," in *A Peep at Washoe* and *Washoe Revisited*, demonstrates the power of illustrations that sensationalized violence in boomtown saloons. Courtesy of the Nevada Historical Society

brought western saloons into American households at least once a week between the 1950s and the 1970s, enhancing the widespread mythic notoriety of these public drinking places. For example, in any given *Bonanza* episode, characters inevitably ran into some sort of trouble while sidling up to the bar in establishments such as the Silver Dollar, Julia's Palace, or the Bucket of Blood Saloon. Fictitiously set in Virginia City, Nevada, this popular series revived the public's fascination with this northern Nevada mining town.[1]

Modern dramatizations, such as those of the *Bonanza* episodes, conveniently coincide with historical accounts of saloons in Virginia City. The original boomtown provided its own backdrop for sensational portrayals of these businesses as early as the 1860s. At that time, Samuel Clemens lived in Virginia City. He and other writers such as William Wright took pen names, such as Mark Twain and Dan DeQuille, and wrote vibrant accounts of saloon shenanigans.[2] Journalist J. Ross Browne published illustrations, such as "Home for the Boys," in prominent magazines like *Harper's Weekly* and *Frank Leslie's Illustrated Newspaper* (figure 1.1).[3] The images depicted by people like Browne and Twain became many people's first impression of western saloons, influencing the violent imagery that brought about later, far-fetched treatments of this social institution.

Brimming with fortune and excitement in 1865, Virginia City provided ample fodder for these writers and journalists. Gold and silver mines yielded enough wealth to transform a camp of scattered tents into a booming city on a rugged mountain slope. Ornate brick buildings and row houses made the place look like San Francisco, while carriages, stagecoaches, pedestrians, horses, oxen, dogs, and even camels filled the dusty streets with a throng of traffic, a clamor of sound, and an overwhelming mix of odors. People going about their business came from all over the world, and street chatter included many foreign languages and heavy accents.

The businesses that lined the bustling cosmopolitan city center included general stores, apothecaries, butcher shops, tailors, boardinghouses, hotels, bowling alleys, and saloons. Saloons were quite common on Virginia City's sprawling urban landscape, and they usually outnumbered all other retail establishments in mining boomtowns. Eliot Lord's estimation of more than a hundred saloons operating in and around Virginia City during the 1870s strongly suggests that they likely outnumbered many other business enterprises in that community as well.[4]

Numerous advertisements in contemporary newspapers portray the assortment of saloons, including those that offered customers billiards, Havana cigars, female entertainment, dancing, coffee, cockfights, dogfights, and a range of alcoholic beverages.[5] This variety suggests several things. First, it indicates that sensational period depictions were based on the sheer abundance of such drinking establishments. Second, shrewd entrepreneurs were trying to fill niches in a saturated boomtown industry. Third, there was a market for a range of businesses devoted to serving the population's need for recreation and leisure. Well-paid miners worked hard underground during eight-hour shifts.[6] They tended to have disposable income, and because of a shorter-than-usual workday, they had free time, especially during mining bonanzas. Saloons served as places where they could while away free time and spend money.

As boomtowns such as Virginia City expanded and became internationally famous, more people arrived from all over the world, amplifying the cultural and ethnic diversity of these communities.[7] Well-established cities therefore supported saloons that filled additional entertainment niches by catering to specific ethnic affiliations. Saloons came to reflect the diversity of these and other urban American centers more than any other social institution. As the diversity of the people increased, so did the need for drinking houses to cater to each group.[8]

Upon arrival in the region's bustling boomtowns, immigrants frequently found a foreign and often hostile environment. The diversity of the people living, working, and socializing in such surroundings meant that the settings could be both intimidating and confusing.[9] Saloons owned by a specific ethnic or cultural group accommodated customers of similar backgrounds and provided places of refuge and solidarity. Although some ethnic groups identified themselves according to places of drink, this was not always the case. For example, even though some Irish establishments in places like Virginia City did not welcome outsiders, other Irish saloonkeepers sought as large a customer base as possible. Another example of diversity was that of a German beer garden in Virginia City, designed by an American-born man named van Bokkelin, who welcomed customers from a broad segment of the community.[10]

Amid this varied collection of saloons, archaeologists have explored—and virtually opened the doors to—four public drinking establishments that

Fig. 1.2. Archaeology crews excavated the sites of the four nineteenth-century Virginia City saloons shown in the context of a thriving boomtown as depicted in Augustus Koch's "Bird's-eye View of Virginia City." Courtesy of Special Collections at the University of Nevada, Reno's Getchell Library

served relatively distinct groups in Virginia City (figure 1.2). The first, Piper's Old Corner Bar, operated between the 1860s and 1880s in the heart of the city's commercial corridor. A German immigrant, owner John Piper opened his business for a range of European immigrants and European Americans. As early as 1861, this saloon operated in a small wooden structure at the southwest corner of B and Union Streets (figure 1.3). In 1863 John Piper built a two-story fireproof brick building across the street, at the northwest corner of B and Union Streets, and it became known as the Piper Business Block. Piper moved the Old Corner Bar to the lower story of that building and rented out the second story as office space. By 1877 Piper remodeled the

FIG. 1.3. Grafton T. Brown's 1861 "Bird's-eye View of Virginia City" illustrates the front facade of the first version of Piper's Old Corner Bar, when it operated at the southwest corner of B and Union Streets. Courtesy of the Library of Congress

business block to become the front section of his newly constructed Piper's Opera House.[11] The opera house had previously operated at another location two blocks downslope from the Piper Business Block, near the corner of D and Union Streets.

Despite the fact that a massive opera house auditorium loomed behind the transformed business block, John Piper still operated his Old Corner Bar from the first story of that building (see figure 2.2). At this time the Old Corner Bar became known as the "theatre saloon," catering to a theatergoing clientele and sporting an upscale atmosphere. The opera house and its saloon thrived as tandem entertainment venues until 1883, when a devastating fire destroyed the auditorium and cleared away the contents of the brick business block.[12] Piper reconstructed the opera house auditorium by 1885, but there is no mention of his reopening the Old Corner Bar at that time.[13] Nevertheless, the outer shell of the brick business block survived the fire of 1883, and the space still housed the charred remains of the Old Corner Bar even after the new opera house rose from the ashes of the old one. Archaeological remains of the Old Corner Bar's 1863–1883 operations lay protected within the rubble of the burned-out shell of the empty saloon space, thus creating a unique archaeological site within a standing structure (figure 1.4).

Piper's Old Corner Bar and Piper's Opera House are relatively well documented in historical records, especially newspapers. These sources help to chronicle the history of the theatre saloon in Virginia City. Documentary evidence about other saloons, however, is not so abundant, and consequently the historical overviews of those establishments are notably shorter.

For example, much less is known about the story of the Boston Saloon. William A. G. Brown, an African American from Massachusetts, owned the

FIG. 1.4. The empty space beneath Piper's Opera House came to life as an archaeology crew unearthed evidence in the darkness of Piper's Old Corner Bar.

establishment and welcomed a clientele who had African ancestry. Brown arrived in Virginia City by 1863, at which time he worked as a "bootblack," a term used for a street shoe polisher. By 1864 he founded the Boston Saloon on B Street, an upslope location along Virginia City's mountainside setting and well beyond the center of town. Sometime between 1864 and 1866, Brown moved his business from that location to the southwest corner of D and Union Streets (figure 1.5),[14] in the heart of Virginia City's commercial corridor and entertainment district. Brown operated the Boston Saloon on that bustling corner until 1875, at which time the establishment disappeared from historical records.

Sources described the Boston Saloon as "the popular resort of many of the colored population," and African American writers lamented the loss of "a place of recreation of our own" in Virginia City after the establishment

FIG. 1.5. This photo of D Street, taken between the late 1860s and the early 1870s, shows the locale surrounding the Boston Saloon. The saloon, to the left of the image, is not visible, but the original location of Piper's Opera House is visible in the second building on the right. Courtesy of the Bancroft Library, University of California, Berkeley

closed.[15] The wording in the sources noted above suggests that the Boston Saloon was primarily for people of African background, and it is quite likely that the drinking house served various socioeconomic segments of that group.

O'Brien and Costello's Saloon and Shooting Gallery operated during the 1870s in the center of the Barbary Coast, a place known for its cribs, brothels, rough saloons, and danger.[16] Owned by men with Irish names, this saloon catered to a primarily Irish clientele, but very likely served other groups as well. While many Irish-owned saloons in other U.S. cities specifically accommodated the Irish, others were more assimilated and sought general patronage.[17] The Shooting Gallery's location in the disreputable Barbary Coast area also suggests that its patrons had little socioeconomic power.

The Hibernia Brewery operated around 1880 just outside the Barbary Coast. This location was about 200 yards from the center of Virginia City's dense commercial district. Typical of the Shooting Gallery and other plebeian establishments situated along the outskirts of the city's commercial corridor, the Hibernia welcomed Virginia City's Irish immigrants and Irish Americans. The Hibernia was owned by two men, Shanahan and O'Connor, whose names evoke Irish ethnicity.[18] The partnership of these two men, like that of O'Brien and Costello, probably allowed the two Irish-owned saloons to stay open all night as the associates in each drinking house shared shifts to cater to the boomtown's twenty-four-hour schedule.[19]

The two Barbary Coast saloons represent the more rustic establishments on the wide-ranging scale of Virginia City saloons. Piper's Old Corner Bar, on the other hand, with its theatergoing clientele, was fancier. Together with Piper's Opera House, this upscale entertainment venue was a good example of Virginia City's transition to a permanent and successful community.[20] The Boston Saloon operated a few blocks downslope from Piper's Old Corner Bar. It was unusual that the Boston Saloon's proprietor was African American, for typical saloon owners were white men in their thirties.[21] That the

FIG. 1.6. Unidentified Virginia City saloon that sported a relatively classy atmosphere and was likely one of the nicer drinking houses. Courtesy of the Nevada Historical Society

Boston Saloon primarily accommodated an African American clientele provides another level of insight into the intricacies that existed among Virginia City's boomtown saloons.

The mere existence of such a diverse range of establishments draws attention to the variety of businesses and people making up the backdrop of mining boomtowns and provides a contrast to the one-dimensional saloon imagery supplied by Hollywood depictions. Historical records portray the more authentic assortment of saloons, with some described as "spacious rooms furnished with walnut counters, massive mirrors, and glittering rows of decanters" while other are characterized as consisting of no more than a "cheap pine bar with its few black bottles."[22] Virginia City's diverse population made for ample clientele to support both the classier establishments and the more sordid ones (figure 1.6).[23]

Together, historical and archaeological records make it possible to go back in time and walk into those establishments. Primary sources aid such reconstructions. Eliot Lord's *Comstock Mining and Miners* (1880), William Wright's *The Big Bonanza* (1876), and nineteenth-century newspapers such as the *Territorial Enterprise,* along with the satirical depictions in Mark Twain's *Roughing It* (1873), are among the records of Virginia City's past written by people who experienced its nineteenth-century mining bonanza. Twain's accounts are certainly fanciful and portray a rather notorious tenure in Virginia City. Twain, like later movies, popularized and mystified the so-called "wildness" of the West. Nevertheless, since he resided in Virginia City during the early 1860s, his descriptions can be used as a type of primary source.

When he was still using the name Samuel Clemens, Twain accompanied his brother, Orion, to Nevada Territory in 1861. Rewarded for supporting Abraham Lincoln in the 1860 presidential election, Orion Clemens received an appointment to serve as secretary to Nevada's territorial governor. Sam Clemens traveled with his brother under the title "secretary to the Secretary."[24] Although he never really held that position because of lack of funding, the younger Clemens remained in the region for nearly three years, joining the staff of Virginia City's *Territorial Enterprise* in 1862. Clemens first used the pseudonym "Mark Twain" in 1863, while he was living in Virginia City.[25]

In addition to primary sources, secondary historical literature, such as the work of Ronald M. James, provides an impressive collection of information about the history of Virginia City. Using statistical data compiled from a wide range of historical records, James assembled an account of nineteenth- and twentieth-century life in this boomtown, presenting a glimpse of its social setting and creating a foundation for understanding the context of saloons in that place.

James teamed up with archaeologists such as Don Hardesty to discover materials associated with the ruins of saloons. Realizing that authentic renderings of saloons had long been clouded—and nearly lost—by popular accounts, these history and archaeology teams were eager to investigate primary historical documents and to sink their trowels into the earth to learn more about these icons of Americana.

Fortunately, "saloon archaeology" in Virginia City stood on the shoulders of many historians who established a scholarly context for the more mainstream dramatizations of public drinking houses. Elliott West's seminal work on western saloons, *The Saloon on the Rocky Mountain Mining Frontier*, provides a detailed compilation of historical documents and images focusing on saloons in the entire Rocky Mountain region. West credits western films, television shows, and novels with introducing the public to the significance of the saloon in the history of the West and observes how historians had, at the time of his book's publication, given these public drinking places "only passing attention."[26] West clearly recognizes the blurred boundary between saloon legend and reality, observing that even though authentic drinking places displayed differences from those of myth, their prominence in the mainstream vision of the frontier is justified on the basis of their sheer abundance and because of their integral role as a social institution in western mining communities. West's analysis emphasizes the social impacts of the mining environment, noting how the society of mining towns in the Rocky Mountain region featured a high degree of alcohol consumption, attributed to the region's early population of many single, lonely men who worked hard, with high ambitions of wealth, yet who were rarely rewarded with such wealth.[27]

While Elliott West's work provides a sturdy foundation for examining saloons in the West, other scholars contribute a broader understanding of saloons in general and a deeper knowledge of these establishments in Virginia City.[28] While saloon studies come from historians, cooperative ventures

FIG. 1.7. The archaeological excavation of the Boston Saloon, shown above at the southwest corner of D and Union Streets, represented the culmination of years of research by historians and archaeologists in Virginia City, Nevada.

between historians and archaeologists have characterized research on these businesses in this particular boomtown.[29]

The synthesizing of historical records and archaeological discoveries is a discipline known as historical archaeology (figure 1.7). While digging is the common means of acquiring actual archaeological artifacts, historical archaeologists dig through libraries and archives and work with historians to determine why they will excavate the earth in search of buried artifacts. Historical records also provide archaeologists with *X*-marks-the-spot maps to guide the placement of excavations.

The techniques of archaeological excavation are often confused with the popular recreational activity known as bottle digging. While both forms of digging provide a means of literally touching the past, they are actually quite different. Bottle diggers search for bottles or other trinkets from bygone days. Archaeologists, on the other hand, search for information. By paying attention to the horizontal and vertical locations of the artifacts that they excavate, archaeologists try to piece together a complete picture of days past. Bottle diggers do not. By digging for information, archaeologists conduct another form of historical research, adding to primary documents. In this way, historical archaeologists bridge the gulf between documents and artifacts, between books and bottles, to refine stories about the past. In this case, they enhance the history of saloons.

Historical archaeological excavations of the four saloons in Virginia City yielded numerous artifacts, making it possible to reconstruct a virtual picture of what it was like to be a consumer, strolling down the various Virginia City streets toward each of the four saloons, approaching their doors, and then walking inside.[30] Such reconstructions add depth to the lively but one-dimensional Hollywood portrayals, introducing a virtual picture of saloons' past.

Piper's Old Corner Bar

It was a brisk January evening in the year 1877. From all directions, winds slashed across the slope of Mount Davidson, and shadows of tree branches danced on the densely packed Virginia City roofs and the steep, snow-covered streets. A well-dressed gentleman ambled along B Street with his head bent slightly against the cold wind and his boots crunching through the icy snow underfoot. He passed under the statue of an unblinded Justice at the new Storey County Courthouse and approached the brick business block that held Piper's Old Corner Bar, the auditorium shape of Piper's Opera House looming behind it. Emanating from three sets of glass double doors, the saloon's gaslights cast their rectangular beams onto B Street. Instead of entering through those doors, however, the gentleman moved toward a fourth, side entrance as he crossed Union Street. Just as ornate as the others, the tall, glass-arched side entry revealed a small group of men gathered at the bar and enveloped by the haze of tobacco smoke. The muffled sounds of piano music and conversation became louder as he drew nearer and reached for the porcelain doorknob.

The warmth of the saloon met a blast of cold wintry air as the storm tried to tug the door from the gentleman's grasp. After a few moments' scuffle, the wind let go and slammed the door shut behind him. His blustery arrival triggered a small windstorm throughout the saloon, causing the chandelier to chime and whisking playing cards out of the hands of two men hovered over an ornately carved stone cribbage board. Sheets of music blew off the piano, but the musician did not seem to notice and kept on playing. One man standing at the bar spat tobacco downward, where it fell squarely into one of the three brown-and-yellow-marbled ceramic spittoons that sat on the floor along the bar. The gentleman nodded at the other patrons, removed his coat and hat, and hung them next to the heavy garments on a series of iron hooks lining the wall near the door.

Recognizing the newly arrived patron, the smartly dressed man behind the bar began pouring brandy into a crystal snifter and pulled out a cigar. He set these at one end of the bar near a glass case filled with coral and seashells. The barkeeper's voice carried a heavy German accent as he greeted the gentleman while clenching the stem of a large meerschaum pipe between his teeth. Free of his snow-dusted outer garments, the gentleman responded to the German's greeting, betraying his own Midwestern American accent. Before sidling up to the bar, he strode across the polished wood floor to a decorative cast-iron stove in the corner, where he rubbed his hands together to warm them.

As his hands thawed, he took in the surrounding scene. A blue haze of tobacco smoke thickened the air. He inhaled the intricate mix of clove cigars, tobacco pipes, spilled ale, and men's cologne. Beyond the smoke, burgundy velvet wallpaper with gold leaf accents testified to the saloon's posh decor. A white ceramic platter held slices of cold beef and mutton, and a tall, hourglass-shaped ceramic decanter sat nearby on the bar. An embossed lion and unicorn reared up to meet each other near the top of this orange-and-tan-colored object, floral patterns and banners decorated its body, and a faucet protruded from its base. The fixture contained a water filter and provided an elaborate means of ensuring that the bar's drinks were not mixed with tainted water. As another antidote to the mining town's poor water quality, the bartender sipped from a glass of mineral water, a product that he imported in abundance from his German homeland. The orange-and-tan-colored stoneware mineral water jugs lined the back bar along with glittering decanters, green glass bottles full of red wine, and a collection of crystal stemware.

The gentleman finally felt too warm after standing so close to the iron stove, so he moved away and joined the other customers telling stories at the bar. His brandy and cigar awaited him, comforts imported from far away to this saloon in the remote desert mountains of northern Nevada.

The Boston Saloon

Snowy, cold winters gave way to mucky springs in the high desert. Lifting her skirt hem above the mud puddles, Amanda Payne gracefully picked her way along C Street one afternoon in 1872. The night before, a spring rain had turned Virginia City's dusty streets into a quagmire, but under the late-afternoon sun the puddles were rapidly evaporating, leaving ruts in their place. The evening shower had also refreshed the surrounding desert hills with splashes of green vegetation and had finally given way to a more usual northern Nevada day with its deep blue, cloudless sky.

Tightening her bonnet, Amanda squinted in the brightness and greeted shopkeepers along the way. A successful African American businesswoman, she was well known among the merchants. She ran a boardinghouse on C Street and was planning to open a saloon and restaurant. From C Street, she gingerly made her way down Union Street's steep slope to D Street. She could feel the rumble of the Virginia and Truckee Railroad down the slope, and she inhaled the aroma of freshly ground coffee from Marioni Brothers

Grocery at the train station. Turning away from the cribs lining the red-light district, she detected the scents of fresh hay, manure, and draft animals from a livery stable at the corner of D and Union Streets. The sound of a trombone playing emerged from the Boston Saloon's open door, adding to the D Street atmosphere and becoming louder as Amanda approached that social establishment.

Although the saloon was brightly lit by gaslights, it took a few moments for Amanda's eyes to adjust from the intense brightness of the sunny afternoon to the interior lighting of the drinking house. Sitting on a chair in the back corner of the saloon, the trombonist filled the small building with his tune. Boston Saloon proprietor William Brown gave Amanda a warm greeting from behind the bar, where he stood polishing glassware. He was helping Amanda set up her saloon, and she had dropped by for a meeting about this matter. She felt a sense of refuge here, as most of the other patrons were also African Americans. Even though she was a respected member of Virginia City's commerce, she still felt the effects of subtle racism. Here, in the Boston Saloon, such subtleties were annulled. She heard the murmured conversation, the ring of poker chips, and the shuffling of playing cards and glanced toward the table of gamblers. One of them chewed on a white clay pipe stem, while another puffed on a red clay pipe.

Gliding over the wooden floorboards toward the bar, Amanda was quite striking in her fashionable dress with its translucent midnight-blue glass buttons. She noticed two young men sipping gin from small glass tumblers and playing dominoes; she did not know them but had seen them polishing shoes on C Street. Dr. Stephenson sat at the bar, taking in the music and twirling a delicate cordial glass in his hands. Crystal stemware sparkled from a cabinet behind the bar. This cabinet also held a stack of white ceramic dinner plates, testament to the saloon's meal service. As if on cue, William Brown's wife emerged from a back room carrying two such plates piled with expensive cuts of mutton. She greeted Amanda and set the food in front of two carpenters at a table near the bar. One of them doused his meal with a red pepper sauce that assaulted Amanda's nose with a vinegary punch as she walked past. Brown's wife lit her own white clay pipe and came over to join the conversation about Amanda's new saloon.

O'Brien and Costello's Saloon and Shooting Gallery

Hearing the sound of a brass horn from one of the D Street saloons, Irish saloonkeeper O'Brien passed the Boston Saloon while walking uphill on Union toward C Street in 1872. Turning onto C Street, he joined the bustle of the downtown corridor, making his way past equestrians and carriages. He soon came upon the Barbary Coast and heard his saloon long before he could see it. Sounds of muffled gunshots floated from open windows of the two-story brick-and-stone building. Coming closer, he heard whoops from drunken patrons in his establishment, which sat in one of the least-respected locations in Virginia City. He and his business partner, Costello, owned the combination saloon and shooting gallery, and together the two kept it open twenty-four hours a day.

Tobacco smoke billowed from the saloon's C Street–facing windows and front door, causing a haze to hover above the boardwalk. Inside, kerosene lamps and candles gave the establishment a dank ambience, especially in contrast with the high desert's sunny autumn afternoon. The odors of gunpowder, stale whiskey, and sweat whirled amid the tobacco smoke, and every few seconds the earsplitting sound of gunfire thundered through the building. A fancily dressed woman flirted with customers who were taking a break from shooting. As they joked with her, he heard the men's Irish accents.

Crossing the room, O'Brien was pleased to see some patrons chatting at the bar and using the fancy stemware he had acquired. He felt that the few pieces of crystal raised the caliber of his establishment. His partner, Costello, was working the bar, so he walked past the shooting gallery and wandered around the back and into the side alley. There he saw a little girl sitting on the ground, chatting to herself while serving tea in a tiny porcelain cup and saucer to a rosy-cheeked doll. O'Brien paused to contemplate the proximity of the innocent tea party to the drunken, vice-ridden activities of his establishment. Such juxtaposition was the nature of Virginia City.[31]

The Hibernia Brewery

Several years later, by 1880, another child played near the festivities of the Barbary Coast. A young Irish immigrant, a miner, saw the child's music as he strolled along the boardwalk toward the Hibernia Brewery. The immigrant had just left a neighborhood a ways downslope where he was visiting family friends from Ireland. He had arrived in Virginia City a few days earlier and had found a cheap room to rent above a saloon at the south end of C Street. The saloon was owned by men of Irish heritage, and they made sure that he felt secure and welcome while awaiting the arrival of his family. His room was on the second story of the saloon's modest wooden building at the edge of this disreputable district.

Approaching the door of the Hibernia, he stepped down onto the sunken floor. Although full of patrons on the warm summer evening, the saloon seemed quiet in contrast to the nearby Barbary Coast. The furnishings were more sparse than those of some of the more posh establishments in the center of town, but the saloonkeepers at the Hibernia focused on serving drinks and fellow countrymen more than they fretted about interior accoutrements.

Kerosene lamps hung on the walls, casting dim beams of yellow light around the perimeter of the saloon. One of the proprietors, Mr. Shanahan, pounded a hammer against the wall, securing a small brass object in place as a decoration near the bar. The object, a bale seal with a lyre design from Ireland, symbolized the nationalistic identity of Mr. Shanahan, his business partner, O'Connor, and the Hibernia Brewery. They stocked the bar with dark green (so dark they looked black) glass bottles and plain glass tumblers. Many patrons leaned against the bar within a fog of their own tobacco smoke. Some chewed on white clay pipes, while others dined on cheap cuts of mutton and pig's feet. A group of men sat huddled around a table playing a game with dice made of bone. All around, he heard Irish accents and the American accents of Irishmen born in North America.

Noticing his fellow countryman and newcomer, Shanahan put down the hammer, poured the young man a draft beer, and raised it in invitation and as a gift. Feeling welcome and safe from the bustle of the cosmopolitan city, the immigrant nodded appreciatively and moved toward the bar instead of retreating to his room upstairs.

THESE VIGNETTES are based on true characters and the objects they left behind. Such artifacts that reflect the material world of the four Virginia City saloons and help us make contact with those establishments. For the most part, the messages that archaeological remains provide from long ago were not intentionally left as keys to history, and because of that they allow us a glimpse into the past of people engaged in their everyday lives—uncensored, unaware, and seemingly honest.

In the effort to understand the bigger picture of saloons, artifacts and historical records provide clues about common activities shared by the various establishments. For example, they shed light on architecture, interior decor, menu items, and entertainment. Different types of materials within each of these categories show how each business diverged from or was similar to its counterparts. In general, differences among the various artifacts seem to reflect the socioeconomic standing of each business, with upscale saloons like Piper's Old Corner Bar sporting wallpaper with gold leaf detail and plebeian establishments such as the Hibernia Brewery offering surroundings that were less elaborate. Such distinctions among this handful of Virginia City saloons help broaden an understanding of the diverse atmospheres and identities associated with each establishment. In order to grasp the nuances of diversity among those public drinking places, it is necessary to take a closer look at artifacts from each saloon.

2

FACADES OF PUBLIC DRINKING

Saloon Architecture

By treating architecture as material culture, it is possible to visualize how people initially experienced the exterior facades of each saloon. Architectural artifacts appeared during excavations of the various Virginia City establishments, and each business's building style can be revealed by combining those objects with historical photographs and drawings. Building materials included various combinations of brick, stone, and wood. In most cases, saloons shared buildings with other businesses, such as a boardinghouse or a mineral survey office. The fact that drinking houses did not always exist as stand-alone structures draws attention to the density of activities taking place within this boomtown's urban landscape.

Whether they operated as single businesses or were part of a montage of other activities, saloon buildings were fixtures in a mining community. Western boomtown architecture reflected both fashionable and vernacular styles that evolved along with the community's transition from mining camp to town, from town to city.[1] For example, tents or shacks characterized the built environment during the first phase of a camp's development (figure 2.1). As a matter of fact, the "first house" that sprang up in Virginia City was a canvas structure that operated as a combined boardinghouse and saloon. A gold mining sluice box served as the bar, and the proprietor sold patrons whiskey but told them they were drinking brandy.[2] By the final stage of a mining camp's maturity, business owners began to invest more time and money in

FIG. 2.1. An early canvas-topped saloon on the Comstock Mining District as illustrated by J. Ross Browne in *A Peep at Washoe* and *Washoe Revisited*. Courtesy of the Nevada State Library

maintaining their establishments and keeping pace with the growing community. Like the rest of a boomtown's commerce at that stage, saloon "architecture" evolved from the rudimentary canvas tents and took on a more refined and permanent appearance. This transformation signaled the shift of the community from a boomtown to a "city."[3]

By the late 1860s Virginia City clearly demonstrated the final stage of community growth, with refined architectural improvements and extravagant interior saloon decor rising from the rough-hewn, nascent camp previously known as "Virginia." While many of the community's drinking establishments certainly maintained rugged appearances at that time, others flaunted opulent architecture, with tall, arched doors and windows and sturdy brick construction, in contrast to the rough-hewn saloon construction made popular by western films.[4] The four saloons highlighted in this study were operating when Virginia City had already established its permanent phase of development and when drinking houses underwent various improvements.

General Architecture of the Four Virginia City Saloons

Piper's Old Corner Bar started out as a small one-story building (see figure 1.3). By 1863 it moved across the street and became part of a sturdy brick block of businesses, known as the Piper Business Block, at the corner of B and Union Streets. Archaeological excavations took place within and alongside the building's intact brick walls. Archaeology crews were able to tell how far down the courses of brick extended, data that provided information about the extent of the building's foundation. That foundation was made up of only two to three courses of bricks buried in concrete-like clay sediments common to Virginia City. Archaeological excavations also revealed a U-shaped trench within the Old Corner Bar space, signaling the footprint of an earlier building at that site.

Eventually, the ornate Piper Business Block became the frontispiece for Piper's Opera House. By 1877 John Piper moved his opera house to B Street, where it overshadowed the business block as a single massive structure. The Old Corner Bar continued to operate from the southernmost space of the first story, making its location literally the southern "corner" of the huge brick building (figure 2.2). The building's east elevation displayed a row of tall windows along the second story, matched by a row of recessed double doors along the first story. Horizontal wood siding, most of which is still visible on the building today, lined the south, west, and north sides of the opera house and business block. The east, or B Street–facing side of the business block maintained its lightly painted brick face. A balcony wrapped around the second story, creating a covered walkway for people walking along B Street in front of the building. Signs indicated the block's activities, advertising commerce such as the saloon operation and the U.S. deputy mineral surveyor.

While historical photos and the fact that the Piper's building is still standing make for an ample understanding of the Old Corner Bar's architecture,

FIG. 2.2. Piper's Opera House, between 1877 and 1880. Note the open doors of Piper's Old Corner Bar at the lower-left corner of the building. Courtesy of Don McBride, the McBride Collection, Virginia City, Nevada

people documented the Boston Saloon less frequently. For that reason, an understanding of that establishment's architecture must rely on the subjective interpretations of a bird's-eye view and scant archaeological remains of a site that was not the first location of the saloon. The Boston Saloon first operated at the north end of B Street, and writer Dan DeQuille, while working for the *Territorial Enterprise,* noted the "suspicious intimacy between Twain and a nigger [*sic*] saloon keeper," describing the establishment in rather negative terms as "a dead-fall on North B Street."[5]

On the one hand, such a description suggests that Brown's first establishment lay at the modest end of the boomtown's scale of architectural improvements, as other places such as Piper's Old Corner Bar had already upgraded. On the other hand, the writer was perhaps unfairly biased by some of Mark Twain's mischief associated with the African American entrepreneur's first business: Apparently, Twain's friends, including Dan DeQuille, accused him of stealing a coat and boots and trading the apparel to an African American saloon owner, presumably William Brown, for a bottle of "vile" whiskey.

Even though this obscure reference provides the fodder for a significant story about the relationship between William Brown and Mark Twain, historical records do not reveal any further information about this matter or about the architectural details of Brown's B Street operation. Twain had moved out of Virginia City by the time William Brown relocated his saloon to D Street, although he might have visited when he returned briefly in 1866 and 1868.

The 1875 bird's-eye view is the only known image of the building that housed Brown's saloon at the new location, 4 South D Street (figure 2.3; see also figure 1.2). This image shows the Boston Saloon as a long, narrow building wedged between several other buildings, with an alley in the back. The densely packed building construction serves as a reminder of the crowded cityscape at the corner of D and Union Streets, a stretch of the community that hosted a constant stream of people going about their downtown business and pleasure.

The building was rectangular, with its long axis perpendicular to D Street. At the center of the Boston Saloon's facade, two windows flanked a door. An overhang or awning spanned the entire width of the facade. The width of the building was no more than 10 or 15 feet, and the length appears to have been about 40 feet.[6] It is not possible to determine from the bird's-eye view whether there were additional windows or doors along the sides of the building or the back/west elevation. Nor can the exact type of siding that covered the structure be determined, although most of the buildings surrounding it on D Street exhibited wood horizontal siding, as can be seen in figure 1.5.

Excavations at O'Brien and Costello's Saloon and Shooting Gallery tell another story about saloon architecture, especially when considered in con-

FIG. 2.3. A modified section from Augustus Koch's 1875 "Bird's-eye View of Virginia City" shows the Boston Saloon at the southwest corner of D and Union Streets; Union Street appears as partially shadowed by the saloon's neighboring building. Courtesy of the Library of Congress

cert with a photo of the building (figure 2.4). Taken in 1967, the photo shows a three-story building with horizontal wood siding. Digging at the site of that building, after it had become an empty lot, archaeologists found that the original structure featured rock side walls and brick front and rear walls.[7] Brick facades on the front and back elevations of the structure conveyed the look of a more opulent structure, while less-expensive building materials made up the building's side walls. Given the dense spacing of buildings in boomtowns like Virginia City, this "deceptive construction technique" was likely a common feature of the urban mining landscape.[8]

The 1967 snapshot is somewhat deceptive as a means of examining the building, since many alterations likely took place between the 1870s when O'Brien and Costello's was in operation and the day in 1967 when the photo was taken. There is, however, still valuable information in the photo, such as the fact that the building stood three stories tall. The Saloon and Shooting Gallery operated out of the first floor, and other activities, such as boarding, probably took place on the second and third floors. The photo also shows several double-hung windows lining the north and east elevations of the second story, with at least one window visible on the east elevation's atticlike third story. The windows on the first story are visible only on the front east elevation of the building. It appears that there were no windows along the north and south sides of the first floor.[9] With no windows to let in light along the side walls, this saloon probably had a rather dark ambience. The lack of side windows did, however, likely make a suitable layout for a shooting gallery.

Similar to O'Brien and Costello's establishment, the Hibernia Brewery operated out of the first story of a two-story building at the south end of C Street. Rectangular in plan, the building sat with its long axis perpendicular

FIG. 2.4. The building that once held O'Brien and Costello's Saloon and Shooting Gallery. Not long after this 1967 photo was taken, the building site became an empty lot, adding another archaeological site to Virginia City's modern landscape. Courtesy of the Comstock Historic District Commission, Virginia City, Nevada

to that street. The building spanned about 60 feet in length and was 20 or 30 feet wide.[10] Because the building that once housed the Hibernia Brewery is no longer extant and because graphics of the structure are absent from the historical records, archaeological investigations were the source for most of the architectural elements discovered that were associated with this operation. For example, masons constructed the building of brick with a repeated pattern of four stretchers and one header. It was then painted white. Archaeologists also discovered a four-foot-deep pit dug into the native sediments. A series of uncut stone blocks arranged around the perimeter of the pit served as a foundation for the building's brick walls. Given this construction style, the first floor of the structure sat slightly below street level, which means people who entered the saloon from C Street stepped down into a sunken room.

Doors and Windows

Naturally, doors provided portals by which people entered and exited saloons; however, western films often show victims of fights flung out of shattering windows, completely bypassing the doors as a means of exit.[11] Whether used for exit in place of windows or not, doors do represent a component of architecture that illustrates the contrast between the western saloons of history and those created by powerful Hollywood imagery. For example, Hollywood typically shows open saloon entrances, framed in weathered wood with short, swinging butterfly doors that fling open as gunfighters, outlaws, and lawmen burst in.[12] Even though this design for doors maintains a strong hold on national and international collective recollections of saloons, it was not actually used in Virginia City. Rather, the portals leading into these places were much more diverse and impressive. A jotting in an 1876 issue of the *Virginia Evening Chronicle* addresses the decorative nature of Virginia City saloon doors: "The saloon keepers of Virginia make elegant doors a specialty, and some of them are marvels of the cabinet maker's art."[13]

FIG. 2.5. Three sets of double doors led into Piper's Old Corner Bar space; the glazing was replaced and doors were repaired and repainted as part of a 2003 rehabilitation of the Piper's Opera House building.

The numerous decorated portals leading into Piper's Old Corner Bar give credence to the above quote. Patrons could enter the Old Corner Bar space from the B Street facade of the Piper Business Block using a series of three double doors with recessed entries, topped by full arches (figure 2.5). Each single half of each set of doors was glazed with two vertical windowpanes, one on top of the other. Additional windows filled overhead lights over each set of doors, allowing natural light to stream into the building during the day.[14] When treated as artifacts, these doors provide a unique sense of being in that establishment, revealing that during the daylight hours, sunshine illuminated this saloon. Located on the east elevation of the business block, the B Street openings received the sunrise as it ascended from the desert floor below Virginia City. By the middle of the morning, the saloon was likely at its brightest, with sunlight streaming through the glass entrances at the front. The historical photo of the Old Corner Bar and Piper's Opera House (see figure 2.2) shows these double doors open, which likely made for a well-ventilated, airy atmosphere as well as a well-lit one.

Another entrance, which also had recessed doors topped by full arches, admitted patrons along the south elevation of the business block. This provided a means of entering or exiting the Old Corner Bar from the corner of B and Union Streets. Without doubt, the Old Corner Bar's doors and the surrounding architecture appear to be the most extravagant of all four saloons examined. It is also interesting to note that there are no reports of barroom brawls at this establishment and there are no reports of people being flung through those elegant portals. As mundane as the reality may seem, the doors at Piper's Old Corner Bar merely served the trouble-free, ordinary functions associated with letting patrons pass and with letting in sunlight or fresh air. Nevertheless, their elegant design indicates that they served those functions with panache.

Unlike the entries to the Old Corner Bar, the doors to the Boston Saloon are neither standing nor intact. The front entrance is visible only as a shaded rectangular area on the 1875 bird's-eye view of Virginia City (see figure 2.3). The doorway is situated in the center of the building's front elevation, as noted above. The available archaeological evidence hardly sheds light on any further details of the saloon's entryway, being limited to hinges and porcelain doorknobs. These came from the western portion of the site, denoting the location of a second, "back" door that led into an alley. Two porcelain doorknobs in a brown tortoiseshell design and two door hinges with gold leaf remnants came from this back door location. While these details provide clues about the rather ornate components of the establishment's back doorway, they cannot help to build an exact picture of the door's actual design. Furthermore, their recovery from the area of the back door does not give any information about the establishment's front entryway. This is an unfortunate, unsolvable mystery, given the significance of saloon entryways in Virginia City, but the back door with golden hinges suggests the potential for an impressive front entrance.

Although the front door remnants were not recovered, archaeologists did unearth window glass from the front, or east, elevation of the Boston Saloon, which was the vicinity of the establishment's front door. According to the 1875 bird's-eye view, there appear to be two windows on the building's front elevation. These probably provided the eastern portion of the saloon with major sources of natural light during the morning hours. Chips of yellow paint cling to the window glass fragments recovered from that area. Although the original wording or design is no longer visible on the fragments of glass, these paint remnants indicate that the building that housed the Boston Saloon had some sort of signage displayed in its front windows. It is also possible that the front door had a glazed panel, similar to those at Piper's Old Corner Bar, which contained such signage or decoration.

The 1967 photograph of the derelict building that once held O'Brien and Costello's obscures a proper view of the building's front door. The doorway appears to be situated in a recessed entry in the center of the front, C Street–facing elevation of the building, with tall windows on either side. Like Piper's Old Corner Bar and like the Boston Saloon, these windows certainly permitted morning sunlight to flood the Saloon and Shooting Gallery. As noted above, the lack of windows along the building's first-story side walls prevented natural light from penetrating very deeply into the back, western portion of that business. Unfortunately, archaeological excavations did not clarify any further door or window details at this establishment.

The frustration with the near invisibility of the Shooting Gallery's windows becomes more discouraging with the Hibernia Brewery. The building and the doors are long gone, and no historical images of the Hibernia remain. The archaeology crews did, however, find evidence of a door that was located on the west, or B Street, side of the Hibernia building. They also discovered a remnant staircase leading from the first-story "back" door up to

B Street. Even so, details of this back door and of the saloon's door that faced C Street are still unknown. Window glass fragments with remnant paint in colors such as yellow, red, brown, and black came from the archaeological excavation at the Hibernia, implying that the establishment had some sort of signage or decoration painted on its windows and/or in its door glazing.

GLASSED-IN DOUBLE DOORS, painted window signs, and multistory buildings are just a few of the architectural components associated with Virginia City saloons. The foundations, the brick walls, tall arched windows, elegant doorways, and painted siding provide evidence of Virginia City's permanence during the 1860s through the 1880s. Each place blended with the urban landscape, surrounded by densely packed buildings. Signs and window paint identified these places. Their doors were substantial, on occasion quite ornate, leading patrons into sunlit rooms. In a way, the exterior architectural components reflected the style of each establishment, and each saloon's character became even more distinct as archaeological excavations unearthed interior fixtures unique to the particular place.

3

AUTHENTIC SALOON SETS

Interior Fixtures

Artifacts from all four saloons included materials from flooring, lighting, wall and ceiling treatments, and novelty items. As archaeologists unearthed these objects, they were able to visualize the sights, smells, sounds, and tastes associated with each establishment.[1] By reviving sensory experiences, the artifacts momentarily transported their excavators into the atmospheres of bygone Virginia City saloons.

Flooring

Archaeologists found that the remnant wood flooring in Piper's Old Corner Bar was charred from a fire in the Piper's building. The identifiable floorboard fragments showed tongue-and-groove construction and were made of redwood. Archaeological excavations at the site that once held the Boston Saloon exposed similar charred wood flooring fragments, but some of these were actually still in place (in situ). Burned in Virginia City's Great Fire of 1875, these floorboards represented casualties of that disaster at the Boston Saloon. The burned floor also told archaeologists that some portions of the site contained pristine deposits that had lain untouched and undisturbed since the days when William Brown closed his D Street business. Realizing that they had reached this long-concealed saloon, the dig team treated the floor area as if it were a crime scene, carefully dusting away the ash with their brushes to see what lay atop and beneath the crumbling floorboards.

FIG. 3.1. Exposed charred-wood floorboards from the building that once housed the Boston Saloon; note the melted and disfigured window glass fragments atop the charred flooring (upper right portion of the photo).

While the flooring appeared to be sitting directly on the surface of the ground, in fact brick footings placed at various locations throughout the saloon space likely supported the floor. Although only a small section of the wood remained intact, that material most likely covered the entire saloon floor. As archaeologists exposed the floorboards, they discovered fragments of window glass adhering to them (figure 3.1). The window glass must have shattered during the fire, becoming molten and losing its flat shape. The fragments then cooled while scattered atop the charred wood, conforming to the shape of the tongue-and-groove flooring, resulting in curved and disfigured fragments of glass. Because the window glass fragments still lay in the position in which they had landed when the window collapsed, the archaeology crew had clear evidence that they were the first people to touch this scene since the fire had consumed the building many years ago; a gold coin found on a nearby floorboard exhibited char damage on one side and a clean, unburned surface on its other side—another example of the site's unscathed deposits (figures 3.2 and 3.3).

In addition to telling archaeologists that story, the window glass suggests that at least one window sat along the north wall. Given the density of the buildings at the corner of D and Union Streets as shown in the 1875 bird's-eye view published in that year (see figure 1.2),[2] the window likely looked into the wall of one of the Boston Saloon's neighboring structures, Number 2 South D Street.

Window glass was not the only material fused to the Boston Saloon's remnant flooring. Burned textile fragments adhered to some of the floorboards, suggesting the presence of rugs or carpeting. Unfortunately, because only a small portion of flooring was found intact, it could not be determined whether the entire floor had some sort of covering or whether only the portion that was preserved had been covered by a runner or carpet.

Excavations at O'Brien and Costello's Saloon and Shooting Gallery revealed a floor constructed of milled lumber and littered with numerous .22-caliber shell casings.[3] Like Piper's Old Corner Bar—and perhaps like the Boston Saloon—in the Shooting Gallery one could easily imagine the

Figs. 3.2 and 3.3. 1854 U.S. 2½-dollar gold piece from the deposits associated with the Boston Saloon's charred floorboards; at left (3.2) is the unburned reverse and at right (3.3) is the burned and blackened obverse. Photos by Ronald M. James

creak of floorboards as people walked through the drinking house. Patrons and proprietors also experienced the scent of wood floors soaked with spilled beer and alcohol, an odor that was described by one historical visitor to Virginia City saloons as a "stench rising from the stained and greasy floors."[4] Although it is unknown whether the Shooting Gallery shared this stench with places like Piper's and the Boston Saloon, people in that Barbary Coast business frequently heard and saw shell casings plummeting to the floor, something that was not common in the other two more respectable drinking houses. The constant presence of shell casings probably made walking rather precarious in certain parts of this saloon. And if the possibility of slipping on the floor was not hazardous enough, the Shooting Gallery's combination of firearms and alcohol surely presented another obvious danger.

At the Hibernia Brewery, archaeologists found floorboards lying directly on the surface of the ground. They also discovered yellow-and-brown-linoleum fragments, indicating that this recently invented floor covering had been installed atop the wood base.[5] A man named Frederick Walton patented linoleum in England in 1863, just seventeen years before the Hibernia went into business. To make linoleum, Walton combined linseed oil, resins, drying agents, and pigments and affixed the mixture to a backing of jute fiber. The same ingredients are still used today to make this type of floor covering.[6] The linoleum likely gave the Hibernia an atmosphere different from that of the wood-floored enterprises. While the linoleum may have lacked the luster of a polished wood floor, its smooth surface probably did not take on the smells of spilled alcohol in the same way that the wood flooring did elsewhere. The use of this innovative floor covering in the sparsely decorated, plebeian Hibernia would have made cleaning and maintaining the floors of this saloon much easier. The practical floor treatment of linoleum, however, does not conform to the traditional Hollywood image of wood-floored saloons. It also departs from the floor treatment used in the three other Virginia City saloons and thus reminds us of the subtle distinctions evident among these four Virginia City establishments.

Lighting

Light fixtures constituted another major element of saloon interiors. Although excavations recovered a few brass fragments from kerosene lanterns at the Boston Saloon, archaeologists located many more items that indicated gas lighting at this site. Among the debris from the 1875 fire were long iron pipes and severely corroded cast-iron fixtures (figure 3.4). One sharp-eyed crew member, while cleaning artifacts in the lab, also noticed some words stamped into a tube-shaped copper alloy object. After the desert dust had been brushed away, it became possible to decipher the words as "HINDRICHS & KNOPP" and "PAT DEC 24 1872." One of the convenient aspects of nondescript artifacts like this is the existence of such an "identification tag" in the form of patent information. In this case, the discovery prompted a trip to the library to review microfilm copies of U.S.

FIG. 3.4. The long pipe and attached fixture in this excavation photo were components of the Boston Saloon's gaslight system.

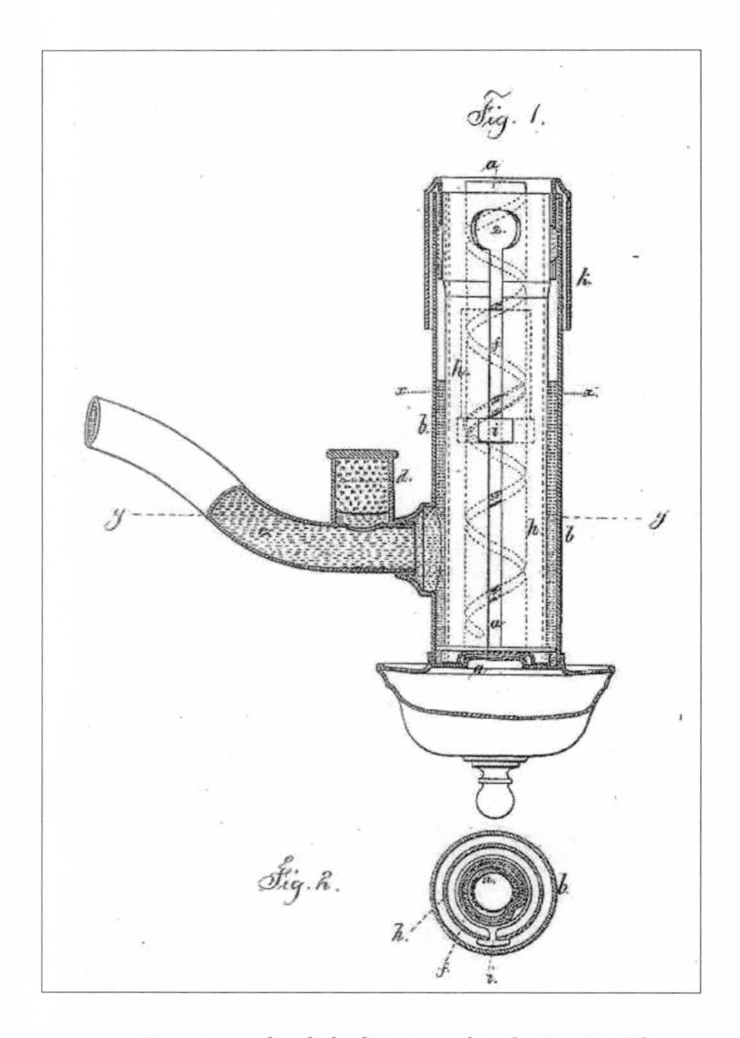

FIG. 3.5. Exact type of gaslight fixture used in the Boston Saloon, as determined by the patent information imprinted on a nondescript metal fragment recovered during archaeological excavations. U.S. Patent Records, Patent No. 134,281, December 24, 1872; courtesy of the Getchell Library, University of Nevada, Reno

patent records. This revealed that the puzzling item was part of a gaslight fixture, patented for an upgrade in gaslight technology that pumped a stream of water through the mechanism as it fed gas to the light fixture, thus supposedly preventing the fuel from generating annoying vapors (figure 3.5).[7]

As at the Boston Saloon, the archaeological record of Piper's Old Corner Bar included kerosene lamps as well as evidence of gas lighting. In fact, pipes from the structure's gaslight system still run along the walls and ceiling of this remarkable site. Decorative brass fixtures and valves provide evidence of the fancy lighting. The gaslights in Piper's and in the Boston Saloon likely imparted a warm yellowish glow to the atmosphere of those establishments. The ambience of their respective interiors certainly emanated from the front windows and doors at nightfall, contributing to Virginia City's evening setting. In addition, the evidence that such lighting systems were used overturns some of the historical negative descriptions of drinking houses as filled with fumes and dimly lit.[8] Despite these reports, the presence of sophisticated gaslights, especially the special patent used at the Boston Saloon, implies that certain bar proprietors attempted to provide their patrons with a cleaner, brighter setting.

Not all saloons had such well-lit interiors, however. The evidence gleaned from the artifacts at the other saloons makes it clear that the light fixtures at the Boston Saloon and Piper's Old Corner Bar set them apart from the boomtown's more dodgy establishments. Since no such artifacts were discovered at the Hibernia Brewery or at O'Brien and Costello's Saloon and Shooting Gallery, it is reasonable to conclude that those two establishments lacked gaslight systems. The excavations at the two Irish-owned saloons recovered fragments from brass kerosene lamps, which may have been mounted along walls, placed on tables, and set on and behind the bar. Even though they did provide light for the interiors of these businesses, it would have been fainter than that provided by gaslights and probably inspired the contemporary characterization of the saloons as "dimly lit." Clearly, the artifacts associated with lighting reveal the subtle ways in which saloon interiors could vary from one place to the next and demonstrate how a nineteenth-century writer could misrepresent the character of all saloons by making generalizations on the basis of observations at certain businesses. If those writers had experienced places like the Boston Saloon or Piper's Old Corner Bar, perhaps they would have described saloons less gloomily.

Walls and Ceilings

Other remnants of the past divulge more obvious interior distinctions between drinking houses, revealing the walls that surrounded patrons while they drank and socialized. Bits of plaster came from all four Virginia City saloons, indicating a common wall treatment, but variations existed among those plaster walls.

Piper's Old Corner Bar had decorative objects affixed to its plaster walls, including a brass star (figure 3.6). This object emerged from the Piper's excavation with plaster remnants fused to one of its sides, giving the impression that it had decorated a wall or ceiling in the building.[9] In addition to

FIG. 3.6. A brass star with plaster affixed to one side from Piper's Old Corner Bar excavations. Photo by Ronald M. James

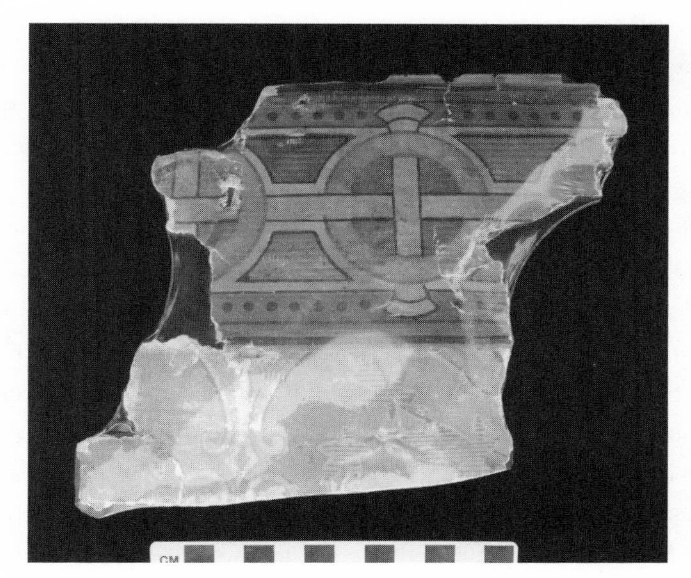

FIG. 3.7. Wallpaper remnants found among the remains of Piper's Old Corner Bar; the fragment at the right is underneath the style shown at the left, revealing the layers of wallpaper used over time. Photo by Ronald M. James

brass ornaments, at least some of the walls in this establishment were wallpapered, representing the fanciest interior element of all four saloons. An entire panel of wallpaper, tattered but still intact, hung along the saloon's north wall. Other scraps of wallpaper clung to various portions of the south wall. At least five different wallpaper designs had been placed on top of one another over time. The designs included an array of colors and textures, with burgundy velvet floral patterns and gold leaf highlights, as well as geometric designs with mixed colors of blue, orange, yellow, and gold leaf details (figure 3.7).

Although the ornate wall treatments in Piper's Old Corner Bar were a tough act to follow, less fancy businesses, such as the Hibernia Brewery, also sought ways to adorn their interiors. Archaeological evidence indicates that the Hibernia's walls were plastered and painted; plaster fragments from that site contained chips of blue, red, gray, white, and black paint. The excavation also found an asphalt-backed fabric adhering to the inside surface of several bricks from the building, which means that, in addition to paint, a fabric wall covering was used on at least one of the Hibernia's walls.

While Piper's Old Corner Bar and the Hibernia Brewery had their distinct wall treatments, the only archaeological evidence about the walls of the Boston Saloon and the Shooting Gallery came in the form of plaster fragments. Despite the general lack of evidence about the interiors of these places, the Boston Saloon excavations did reveal small, thin, decrepit tin fragments with a stippled pattern, which indicated that patrons in this saloon socialized beneath a decorative pressed-tin ceiling. Developed in the middle nineteenth century, such ceiling treatments relied on mass-produced sheets of thin, rolled tinplate and other metal sheets such as copper, which were stamped with complex patterns that mimicked carved and molded plasterwork, such as that used in the wealthiest European households.[10] Although prevalent throughout the latter half of the nineteenth century, the style reached its peak of popularity by the 1890s, which means that the Boston Saloon's use of this embellishment was rather innovative and quite trendy. Such decorative ceilings gave many Virginia City interiors a fashionable flair, and the stippled-tin fragments that emerged from the ruins of the Boston Saloon suggest that it was clearly among the more stylish places in the boomtown.

Novelties, Identity Symbols, and Folk Beliefs

Interior chic was not limited to tin ceilings and velvet wallpaper. As a matter of fact, as patrons shifted their gaze from walls and lights toward the less mundane elements of each saloon, they could discern even more diverse ways in which saloonkeepers flaunted their businesses' status, taste, and overall identity. For example, archaeologists at Piper's Old Corner Bar recovered a distinct novelty when a concentration of coral, seashells, and a crab claw emerged from the collapsed remains of that site (figure 3.8). Shattered flat glass fragments surrounded these items, indicating that some type of glass-walled container, such as a display case or curio, had held the unusual collection.

Although it cannot be proved, it is possible that the oddities from the sea represented the remains of an aquarium. Marine aquariums developed during the 1850s, and the ensuing social craze over them let to their placement in almost any home or enterprise that attempted to boast of affluence and modernity.[11] For the first time in history people viewed the undersea world in cross-section instead of merely looking down into the water and imagining life under the sea. An 1873 article from a Carson City, Nevada, newspaper touted one of these early aquariums in San Francisco: "It is difficult to describe the novelty of the sight. There are three marine tanks already

Fig. 3.8. Aquarium or curio items, including shell, crab claw, and coral fragments. Photo by Ronald M. James

filled—one with crabs, etc.; second with perch, flounders and a variety of small salt-water fish; and a third with star-fish, etc. . . . Our reporter never before realized the beauty of a fish, and the wondrous grace of its motion . . . one also has a good opportunity to watch and study the habits of the fish while close at the bottom of the sea."[12]

Whether Piper's featured an aquarium or not, the extent of unique fixtures at that establishment is quite remarkable in the context of other Virginia City saloon artifacts. The Hibernia Brewery also had its own material distinction, in the form of a brass bale seal decorated with a double-headed dragon forming the body of a lyre (figure 3.9). When recovered during the archaeological excavation, this object appeared to signify the establishment's Irish identity, as it originally sealed a product that was likely shipped out of Ireland. When archaeology teams dug it from the Hibernia's buried remains, they discovered that it had been affixed to a piece of wood with a nail, implying that long ago, someone had nailed the seal to a wall or overhang to create a meaningful, and perhaps nationalistic, interior decoration. Such symbolic re-use of a rather utilitarian object implies Irish ethnicity, or possibly ethnic pride, associated with this saloon.[13] This episode also illustrates the importance of paying attention to the ways that certain everyday objects could serve as symbols of a group's or individual's identity.[14]

While the Irish outwardly boasted of elements of their group identity in places like the Hibernia Brewery, an African American placed at least one small group of objects underground at the Boston Saloon. Near the last day of excavation at that site, crew members exposed a fully articulated small mammal skeleton lying on a flat rock near two U.S. coins, one of them a silver dime minted circa 1853–1860 and the other an 1865 silver half-dollar. These items lay beneath the undisturbed but charred floorboards, indicating that they were concealed from view during the Boston Saloon's operation.

The coins were burned and disfigured, a condition that was initially attributed to the Great Fire of 1875. However, the coins had other modifications that clearly signaled human, or cultural, input (figure 3.10). For example,

FIG. 3.9. The design on this brass bale seal shows a double-headed dragon forming the body of a lyre. The seal, recovered during archaeological excavation at the Hibernia Brewery, appeared to signify the establishment's Irish identity. Photo by Ronald M. James

someone had made a pinhole-size perforation near the edge of the dime, causing a small fracture that ran from the point of puncture to the edge. The words *half-dollar* and the date *1865* were barely visible on the other coin, which had two punch marks that made two trapezoid-shaped perforations in the center of the coin. Immediately beneath these alterations, a narrow cut extended from the lap area of a seated Liberty on the obverse edge of the coin. A third coin, a perforated Chinese disk, emerged during excavation within one meter of these other two modified specimens; however, the hole in that coin may represent an imperfect casting (figure 3.11).

Besides the punctures and cuts, the coins were disfigured by burning, as the metal was probably heated by someone who wanted to soften the coins to prepare them for the various modifications. The fire of 1875 cannot be the

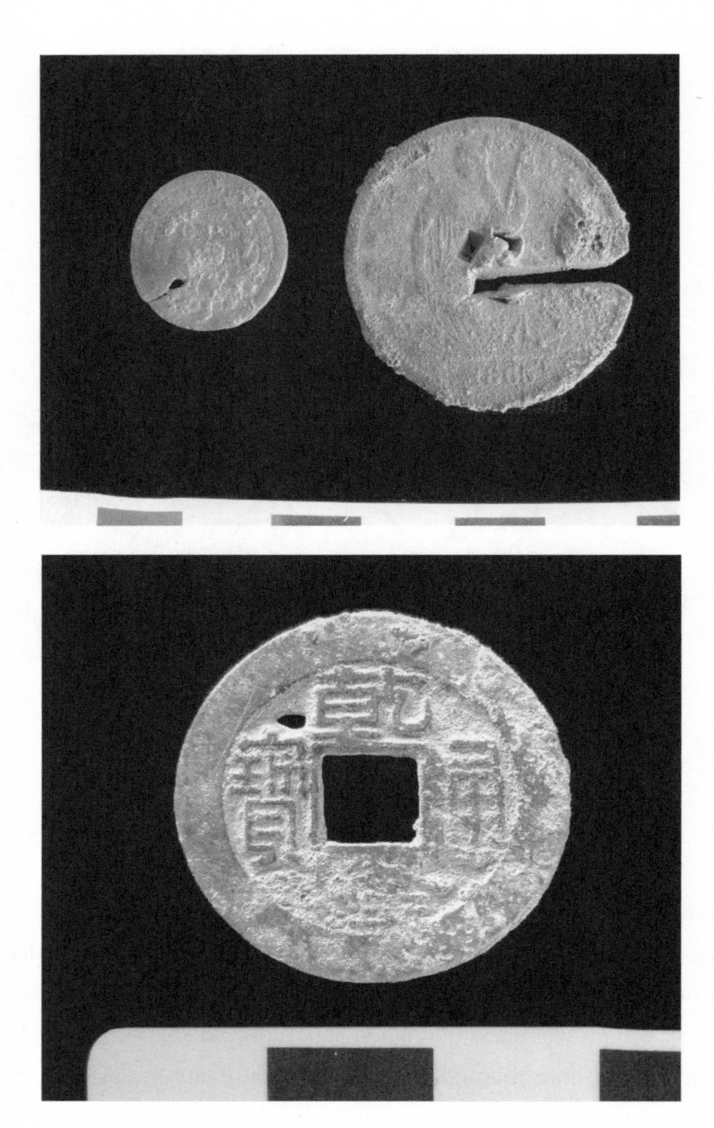

Top: **Fig. 3.10.** These two modified coins from the Boston Saloon evoke nineteenth-century folk beliefs in a Virginia City saloon. Photo by Ronald M. James
Bottom: **Fig. 3.11.** Chinese coin from the Ching Dynasty, 1644–1911, from excavations at the Boston Saloon. It is a Ch'ien Lung period "cash" piece made between 1736 and 1795; circular perforation may represent an alteration similar to those of other modified coins shown above. Photo by Ronald M. James

cause of the coins' disfigurement, because they were hidden—and consequently protected—beneath the floorboards.

Coins are an expected find in a saloon, where money was commonly exchanged for food and beverages. But these modified coins require closer attention, because the appearance of similar coins at other nineteenth-century African American sites has been linked with folk beliefs that they served as charms "charged and imbued with attributes for supernatural control."[15] Furthermore, the fact that archaeologists found no perforated coins at the other Virginia City saloons strongly suggests that they may represent material vestiges of something that had meaning for African Americans in Virginia City. Although archaeologists have linked this use of charms with African Americans elsewhere, the protective status of perforated silver coins is also associated with a European folk belief dating from at least the sixteenth century that was centered on protection from witchcraft and the supernatural.[16]

James Davidson, of the University of Texas at Austin, expanded the complex background of coin use after examining perforated coins described in the narratives of elderly ex-slaves collected during the 1930s by the Works Progress Administration (WPA). Davidson argues that these should not be misconstrued as ethnic markers because silver coins modified as charms can also be traced to European folk beliefs, the British Isles, and even to an English settler in seventeenth-century Jamestown, Virginia. There were still a few occurrences of Europeans carrying on the practice during the middle and latter nineteenth century, such as a group of Swedish immigrants working in the copper mining industry in Michigan's Upper Peninsula.[17] Even so, by the late nineteenth century, African Americans were the predominant users of this form of charms.[18] The folk beliefs associated with modified coins, especially silver ones, have been described as a combination of European charm use and West African beliefs about healing ailments and protection from conjuration.[19] The use of perforated coins possibly represented an active attempt of freed and enslaved African Americans to gain control over at least certain aspects of their lives. Though African Americans often wore the perforated coins around the neck or the ankle, they also buried them beneath building floors in domestic contexts.[20]

The discovery of modified coins in a late-nineteenth-century African American saloon in the mining West inspires a consideration of the persistence of this practice across space and time, from colonial America to the mining West. While the perforated dime recovered from the Boston Saloon is consistent with the types of materials that have been associated with charms and conjuration, the modified half-dollar is an anomaly. In his examination of the WPA ex-slave narratives, Davidson observes that the informants do not discuss any means of altering coins other than perforation.[21] At this point there is no explanation for the strange, cut-like modifications of the half-dollar from the Boston Saloon, but its association with the perforated dime implies that its modifications may reflect similar symbolism.

On their own, the mysterious coins may merely be a coincidence; they cannot unequivocally be interpreted as representing a ritual activity in African American culture. On the other hand, their concealment beneath the Boston Saloon's floorboards adds some credence to that argument. Similar coin deposits discovered at other African American sites were also hidden, a practice that might imply an effort to avoid persecution for employing such folk remedies. African Americans lived in a prejudicial world and did not have the luxury of openly practicing their folk beliefs. To avoid persecution, especially under the conditions of enslavement, they continued these traditions in "underground" contexts.[22] It remains a mystery whether the person(s) who modified and concealed these coins placed them beneath the floorboards to hide their beliefs or merely to make the ritual more effective. Their apparent concealment beneath the floorboards also indicates that they probably had meaning for only the person or persons who placed them there.

This episode, though an unsolved mystery, may nevertheless offer valuable information for archaeologists who are trying to revive the human experience in mining boomtowns, inspiring thoughts about the presence of folk beliefs and superstitions in those communities. As diverse groups of people from all reaches of the globe flocked to places like Virginia City, Nevada, they transported an assortment of cultures and belief systems to that isolated urban center in the rugged northern Nevada desert. Many of those beliefs were certainly grounded in deep-rooted superstitions and fears that could be traced back many generations in the Old World.[23] As a result, a mass of superstitious folklore thrived within the context of Virginia City's daily life. It is impossible for us to get into the minds of the people going about their business under various shrouds of ancient superstition, but artifacts such as the perforated coins provide clues about at least one form of such beliefs. Although these particular kinds of objects were not found in the other Virginia City saloons, many other objects suggestive of superstitious beliefs were unearthed during the archaeological work in this cosmopolitan setting. For example, a cache of objects including a hat, a boot, scored leather, a wine bottle, and a padlock appeared during the excavation of a small house on the outskirts of Virginia City's Chinatown. The association of these items together in that deposit prompted interpretations connected with superstition,[24] which calls attention to the traditions and beliefs that people carried to the American West from throughout the world. But wait, there's more. In August 2004, just before this book went to press, archaeologists digging in Virginia City's Chinatown found a fully articulated small mammal skeleton underneath an intact foundation stone. Whether this represented yet another means of warding off unwanted spirits is unknown, of course; but the results of the subsequent analysis will be intriguing.[25]

Indeed, one of the greatest of the ancient fears transported from the Old World was that supernatural beings would try to enter someone's house to inflict mischief. To ward off such beings, people placed dried cat carcasses or footwear inside the walls of various structures, including houses, churches, and taverns; these practices can be traced back hundreds of years in European history, to at least the 1400s, as a means of keeping "spiritual vermin and witches at bay," and evidence of these items—that is, evidence of superstition—has emerged in nineteenth-century archaeological sites on American soil.[26]

Given the broader context of superstition, the Boston Saloon's perforated coins were analogous to people's hopes that they could improve their chances for good luck or fertility by burying footwear beneath the floors or in the walls or chimneys of their homes or businesses.[27] Another kind of ritual illustrating a rather credulous practice of folk medicine was illustrated by Mary McNair Mathews, one of Virginia City's well-known citizens who was particularly outspoken and venomous toward minority populations. After one of her son's fingers was accidentally cut off, she placed it in a brandy jar as a means of protecting him from future harm.[28] Mining communities such as Virginia City were riddled with other forms of folklore as well, such as the Cornish belief in Tommy-knockers, which were small elfin creatures that live in abandoned parts of mines.[29]

The various folk beliefs reflect Virginia City's bygone days when common objects were imbued with meaning. Laced with intrigue, the Boston Saloon's concealed coins still have the potential to signify the culture and identity of their holders. Such artifacts pose a tantalizing challenge to archaeologists as they work with shreds of commonplace material evidence to make sense of a time when objects were regarded as much more than powerless entities and to account for a global context of superstition in isolated western boomtowns.

Perforated coins, decorative wallpaper, a re-used bale seal, an aquarium showcasing treasures from the sea—all these objects certainly had an effect on how the people who came into their presence felt, and they combined with all the other elements of the saloons' ambience to create each establishment's unique identity.

MENU ITEMS

Drinking and Dining in Virginia City Saloons

Unsurprisingly, archaeological excavations at Virginia City saloons turned up a profusion of bottles and bottle fragments. Old bottles from ghost towns are sought-after collectibles, enticing people to view them as buried keepsakes. When these little conversation pieces are retrieved using meticulous archaeological methods, they can do more than merely sit atop a bottle hunter's mantelpiece. They lead us into the experience of being in a historic saloon by revealing clues about what people drank. Other objects unearthed during the archaeological investigations, although not as alluring as shiny glass bottles, enhance this experience. Animal bones, condiment containers, and trace elements of food residue, along with the various bottles, provide clues to the diverse food and beverage menus of Virginia City's drinking establishments.

Some Virginia City saloons advertised "well kept and assorted lunches," "choice lunches," and "superior hot lunches."[1] Given the relatively large populations of lonely single men in mining boomtowns, saloonkeepers wisely responded to a clientele in need of snacks and meals, as well as drinks. Common meals served at nineteenth-century saloons included cheese, potatoes, and mutton pie, and drinking houses in "humble neighborhoods" offered hot meals at certain times of the day.[2]

While food remained a novelty and a draw for potential patrons who were looking for a good meal, many saloon proprietors advertised their establishment's drink menus more frequently than their food offerings. Some Virginia City ads are general, with such verbiage as "the very choicest wines, liquors and cigars . . . everything first-class," while others are more specific: "sole agents for the following brands of fine old whiskies: William Crowder's Kentucky Bourbon, Club House Favorite, Clipper, Reynold's Rye, Dickson's Old Farm."[3] Of the four Virginia City saloons being considered here, only Piper's Old Corner Bar regularly ran advertisements in local newspapers. For unknown reasons, newspaper advertisements for the others are virtually absent, making the archaeological record the sole "document" that contains clues about the menu items served in those establishments. In Piper's Old Corner Bar ads, the saloon boasts the "finest brands of wines, liquors and cigars, at one bit . . . [with] every attention paid to patrons." In other ads, proprietor John Piper lists specific products, such as "Old Bourbon" and "Cutter's" whiskies, "Napoleon Cabinet" and "Imperial Cabinet" champagne, and "Otard, Dupuy, and Co." brandy.[4]

Intoxicating Beverages

Although labels from these "finest" brands did not appear during the archaeological analysis of bottles from Piper's Old Corner Bar, the numerous wine, champagne, and ale bottles made of dark green glass that were excavated underscore the popularity of those beverages both at Piper's and at the other three establishments. A few intact specimens and thousands of fragments from such bottles illustrated that this type of beverage container was the most common and abundant bottle type represented at all four Virginia City saloons (figure 4.1).[5]

The occurrence of green glass bottles at the four saloons is analogous to the presence of buff-colored pottery, or stoneware, bottles, at all four establishments (figure 4.2). Most of the stoneware bottles display stamped manufacturer marks near their bases, indicating that they contained ale and that they came from Glasgow, Scotland.[6] Items such as wine, champagne, and beer—from green glass and from stoneware bottles—appeared on the

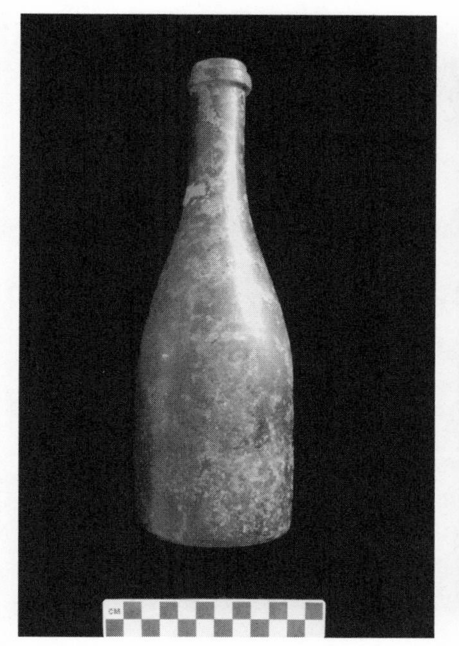

Left: **FIG. 4.1.** Dark green wine and champagne bottles are represented by a few intact specimens, such as this one recovered from the Boston Saloon, and by thousands of glass fragments. Photo by Ronald M. James

Right: **FIG. 4.2.** Stoneware ale bottles and shards, including those with "GLASGOW" stamped near the base, emerged during excavations of the four Virginia City saloons. Archaeologists found the bottles shown here in numerous fragments. Photo by Ronald M. James

menus of various businesses, then, regardless of the socioeconomic and ethnic affiliations of the particular place.

Despite the similarity among beverage containers across the sample of saloons studied, at least one product was unique to the Boston Saloon: "Gordon's Gin London," as revealed by the embossing on rectangular bottle fragments with a light aqua-green tint (figure 4.3). By the late nineteenth century, when the Boston Saloon was operating in Virginia City, gin was in transition as a drink associated with different socioeconomic groups. Its ori-

FIG. 4.3. Archaeology crews discovered this glass fragment from a flat-paneled Gordon's Gin bottle at the Boston Saloon. Photo by Ronald M. James

gins date to seventeenth-century Holland, where the Dutch used it as a medicinal remedy for stomach ailments, gout, and gallstones; eventually they flavored the product with juniper, from which it gets its name. By the end of the seventeenth century the drink spread to England, where it became a preferred drink of the poor. This trend continued for nearly 150 years, with England's lower classes drawn to lavishly furnished "gin palaces." By the 1870s gin became desirable to high-society drinkers; this transition came soon after the gin production process in England became more refined in response to licensing reforms.[7] While the presence of gin bottle fragments could be used to make a statement about the class status of the Boston Saloon's patrons, that would be pure speculation since the business operated on the temporal cusp of gin's socioeconomic shift. Nevertheless, other bits of archaeological evidence from this saloon suggest that it was among Virginia City's finer establishments, which certainly substantiates an explanation of gin's association with "high society" by the time the Boston Saloon opened.

Even though the Gordon's Gin bottle fragments invited archaeological interpretation, it is important to understand that bottles are not necessarily the definitive hallmark of all saloon beverages. An abundance of barrel staves, barrel hoops, faucets, and bibbs—a type of faucet with a bent-down handle—remind us that many beverages, such as whiskey, came out of kegs (figure 4.4).[8] Archaeological excavation recovered significantly more faucets from Piper's Old Corner Bar than from the other three saloons (figure 4.5), likely because Piper's had a fairly large cellar and most of the faucets came from kegs that were stored there.[9] No such cellar or storage feature existed at the other establishments, suggesting that they lacked the space for kegs and the tap mechanisms that accompanied them.

Glass decanter stoppers are among the other items associated with drink service in these public drinking places. Archaeologists recovered three decorative stoppers from Piper's Old Corner Bar and one from the Boston Saloon, implying a level of opulence reminiscent of Eliot Lord's description of the more upscale Comstock saloons as having "spacious rooms . . . and glittering rows of decanters."[10] While the Hibernia excavation revealed no evidence of fancy decanter stoppers, O'Brien and Costello's Saloon and Shooting Gallery yielded at least two specimens, one of which appears to be the fanciest decanter stopper of all four Virginia City saloons. This ornamental colorless glass specimen features an elongated handle decorated with interlocking diamond-shaped facets (figure 4.6).

The occurrence of such an ornate object at a Barbary Coast saloon demonstrates the value of the archaeological record, since it is an item that a researcher would expect to find at a more upscale establishment. Its presence at a place like O'Brien and Costello's provides evidence of the complexities of Virginia City saloons; they did not always neatly fit into categories described as spacious rooms with mirrors or places with cheap pine bars and a few black bottles.[11] While these extremes certainly did exist, many establishments probably combined some characteristics of each extreme, falling somewhere in the middle. Even though O'Brien and Costello's Saloon and

CM

CM

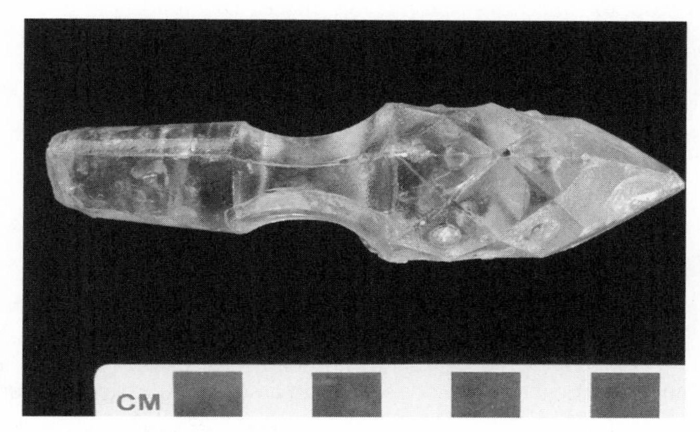

Fig. 4.6. Glass decanter stoppers added flair to O'Brien and Costello's Saloon and Shooting Gallery; this one is the fanciest of all the stoppers found, which is somewhat ironic since O'Brien and Costello's establishment operated in the Barbary Coast, the seediest part of Virginia City. Photo by Ronald M. James

FACING PAGE:
Top: **Fig. 4.4.** Corroded remnants of a faucet like this tapped Boston Saloon kegs. Photo by Ronald M. James
Bottom: **Fig. 4.5.** These faucets came from Piper's Old Corner Bar; the style on the left with the bent-down spout is a "bibb," while the style on the right is a standard faucet often referred to as a "cock." Photo by Ronald M. James

Shooting Gallery was located in a sordid neighborhood, the decorative decanter stopper illustrates that the low status of a saloon did not necessarily prevent its saloonkeepers from using materials associated with higher-status operations.

Other artifacts indicate the presence of pharmaceutical products at the various saloons. For example, the largest quantities of intact bottles unearthed during the Boston Saloon dig were aqua-blue "Essence of Jamaica Ginger" bottles (figure 4.7).[12] Historical newspaper advertisements describe this product, known as white ginger, as a cure for nausea and other "diseases" of the stomach and digestive organs.[13] The abundance of this so-called pharmaceutical remedy at a saloon, in association with numerous alcohol bottle fragments, could indicate a cure for the effects of overindulgence in alcohol. Or perhaps Jamaica Ginger provided a substitute for alcoholic beverages.[14] Ginger might also have been added to ale to make a flavored

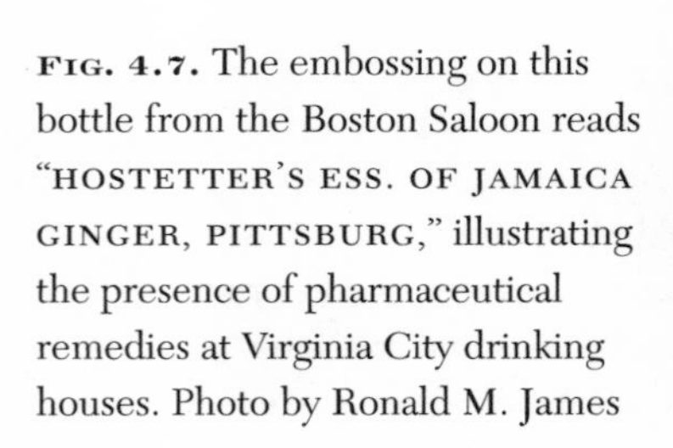

FIG. 4.7. The embossing on this bottle from the Boston Saloon reads "HOSTETTER'S ESS. OF JAMAICA GINGER, PITTSBURG," illustrating the presence of pharmaceutical remedies at Virginia City drinking houses. Photo by Ronald M. James

beer, or it could have been combined with soda water to make a nonalcoholic ginger-flavored drink. Although there is no way to verify the use of this product in such mixtures, there is evidence of soda water bottles and ale bottles at the Boston Saloon, suggesting that all of the above could have been menu options at that establishment.

While "Essence of Jamaica Ginger" appears in quantity at the Boston Saloon, "Dr. Hostetter's Stomach Bitters" was similarly prevalent at Piper's Old Corner Bar. During the nineteenth century, people believed that bitters (spirits of varying alcoholic "strengths" flavored with roots and herbs) cured various ailments, including cholera, malaria, and diarrhea.[15] The Boston Saloon had a unique form of bitters as well, packaged in a green glass container that initially appeared to be a wine bottle. The bottle had a circular seal on its shoulder embossed with the words *FRATELLI/ . . . RANCA/MILANO* encircling a starburst; this mark indicates that the bottle originally contained Italian bitters (figure 4.8). Elsewhere in Nevada's mining districts, such bot-

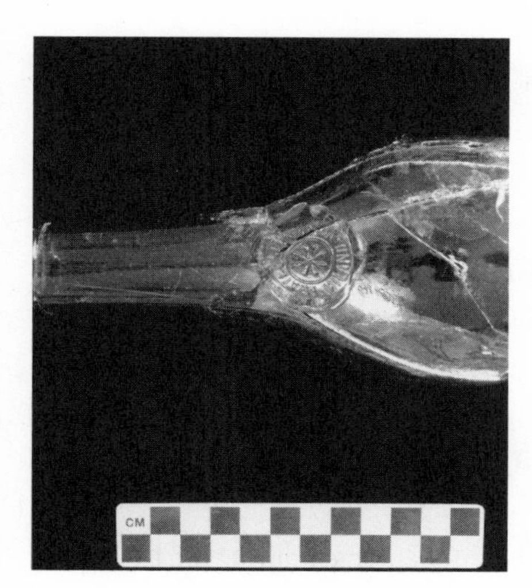

FIG. 4.8. Bottle from the Boston Saloon that once held Italian bitters, as indicated by circular seal on the shoulder, reading "FRATELLI/ . . . RANCA/MILANO." Photo by Ronald M. James

tles have been interpreted as Italian ethnic markers.[16] It is uncertain whether this bottle indicates Italian clientele at the Boston Saloon, but its presence, along with such items as Gordon's Gin and Essence of Jamaica Ginger, calls attention to the diversity of this establishment's menu items.

Bitters and Jamaica Ginger are present in both the Hibernia and O'Brien and Costello's assemblages, but the Boston Saloon has the highest number of Jamaica Ginger bottles and Piper's has the highest concentration of Dr. Hostetter's Stomach Bitters containers. It is entirely possible that these establishments offered such products as digestive aids,[17] and there were occasions where saloons advertised bitters and essences, along with wines and ales, in local newspapers.[18] It was also not uncommon for medicinal products like stomach bitters to be favored by proponents of the temperance movement, who could thereby enjoy a social drink in the saloon atmosphere without compromising their vows and pure intentions or incurring the guilt involved in drinking alcohol. Were they unaware of the fact that bitters usually was relatively high in alcohol content.[19]

The Novelty of Drinkable Water

Some saloons offered other nonalcoholic alternatives, such as artificially carbonated water—that is, soda water. Although it was noted as a popular mixer in Virginia City stores, soda water bottles were relatively scarce in the four saloon archaeological assemblages.[20] A common soda bottle type came from the Boston Saloon and the Hibernia Brewery; its fragments were embossed "CANTRELL & COCHRANE," a company based in Dublin and Belfast (figure 4.9). Sir Henry Cochrane and Dr. Thomas Cantrell opened their soda business in those cities during the 1850s, and their company dominated the Irish market throughout the following century and still operates today.[21]

The commercial production of soda water emerged during the latter portion of the eighteenth century in England. The product's popularity spread quickly through Europe, and by 1807 soda water was available in North

FIG. 4.9. Aqua soda water bottle fragments like this rounded bottle base, embossed with "CANTRELL & COCHRANE," were found at the Boston Saloon and the Hibernia Brewery. The round base ensured that the bottles were stored on their sides, which prevented corks from drying out, shrinking, or being pushed out by contents under pressure. Photo by Ronald M. James

America. It eventually became more popular than mineral water, because it kept its carbonation longer. Along with mineral water, soda water provided remedies for dyspepsia, constipation, fever, and thirst. It began to compete with mineral water as a nonalcoholic temperance alternative and also as a mixer to be used with spirits. Soda water eventually was used to make lemonade and other flavored beverages, a practice that culminated in the modern carbonated beverage industry.[22]

Evidence of another nonalcoholic item, one of soda water's competitors, also appeared at the excavations of all four Virginia City saloons: German mineral water. This product was packaged in tan stoneware bottles, also referred to as "jugs," and could be identified by a stamp on the upper portion of each vessel on the "shoulder" area, opposite the handle (figures 4.10 and 4.11). This circular stamp indicated the well from which the water was harvested and included a crowned eagle at its center. The letters *FR* on the eagle's chest signified "Frederich Rex," or Frederick the King, which helps associate the jug's manufacture with the Prussian Empire.[23] The words

FIG. 4.10. These stoneware mineral water jugs from Piper's Old Corner Bar are the same type as those found at the other Virginia City saloons. Photo by Ronald M. James

SELTERS//NASSAU surrounded the eagle insignia in a ring-shaped band, also called a torus. This mark revealed that the bottle was made for the tenant(s) who leased the well or spring at Selters in the duchy of Nassau, Germany. Nassau is located in a region known for its mineral springs, and *selters* refers to Nieder Selters, a town in this Wiesbaden region from which seltzer water takes its name.[24]

FIG. 4.11. Close-up of stoneware mineral water jug seal, showing the label "SELTERS//NASSAU." Photo by Ronald M. James

Although there were relatively few Selters mineral water jugs at the other saloons, the number of these items at Piper's is staggering; archaeologists at the Old Corner Bar recovered at least eighty.[25] The abundance of these objects at Piper's suggests that John Piper purchased the product in bulk, perhaps buying an entire crate, or "shook."[26] It is possible that Piper, a German immigrant from a region not far from Selters, developed a taste for the water despite reports of its rather salty and unpalatable flavor.[27] It is noteworthy that the quantity of the product is so high at the German-owned saloon and relatively lower at the other establishments. It is difficult, however, to say whether this difference is the result of a taste preference, a health fanaticism, a good deal on buying the product in bulk, or different sampling methods employed in the archaeological excavations at each saloon. What can be said at this point is that some Virginia City saloonkeepers and patrons likely drank the imported mineral water for health purposes, as a temperance-influenced alternative to alcohol, and as an ingredient in the preparation of mixed drinks.[28]

The unique refreshment from the Nassau region, as well as soda water, would have provided occasional alternatives to the region's natural water supply. Numerous writers attested to the poor quality of the local water, among them journalist and artist J. Ross Browne, who arrived in Virginia City in 1860 and became sick from drinking the water there. He conse-

FIG. 4.12. This large decorative water filter graced the interior of Piper's Old Corner Bar. Photo by Ronald M. James

quently reported that "the water was certainly the worst ever used by man."[29] Comstock residents lived with the poor quality of this water from 1859, when the mining district was founded, until the completion of a flume system in 1873.[30] The discovery of such an abundance of German mineral water jugs strongly suggests that Piper's Old Corner Bar used this product to help cope with the unpalatable drinking water during that fourteen-year period.

The archaeological investigations at Piper's Old Corner Bar yielded one other option for enduring the water problem—an elaborate water filter. This decorative stoneware vessel looks like a tall decanter with two side handles, a lid, and a faucet hole (figure 4.12). Encircling the vessel are geometric bands and beaded designs, interspersed with embossed floral and plume decorations and the royal seal of a lion and a unicorn rearing up to meet each other. In addition, there are banners bearing the words *PATENT MOULDED CARBON FILTER // F.H.ATKINS & CO.// 62 FLEET STREET//LONDON.*[31] This device provided purified water for drinking or for mixing with beverages, and its intricate design certainly added a classy element to the ambience of Piper's Old Corner Bar. Archaeologists did not find any other water filters during their excavations of the other Virginia City saloons. The presence of this water filter clearly distinguished the posh establishment of Piper's from Virginia City's less extravagant places of social drinking.

Animal Bones and Saloon Meals

The types of foods served in each saloon provide another level of distinction among the four sites. While digging through the rubble of all of the Virginia City drinking places, archaeologists exposed numerous faunal remains, or animal bones. Cut marks indicated that the animals had been butchered for eating. The presence of the bones provides information about some of the meals offered by the various saloons.

It is possible to identify animal species from certain bones, which helps archaeologists determine what kinds of meats people consumed.[32] The bones of domestic animals were most common among those collected from

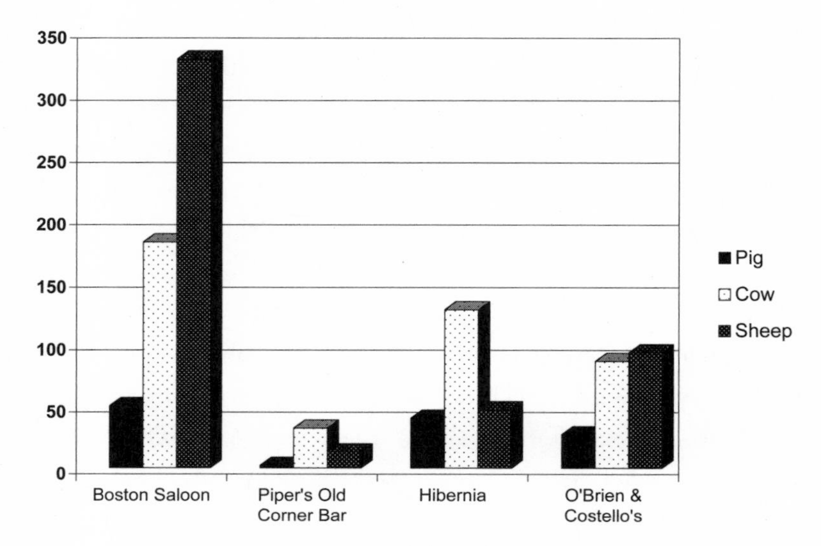

GRAPH 4.1. Summary of pig, cow, and sheep frequencies at the four Virginia City saloons; note the predominance of lamb at the Boston Saloon.

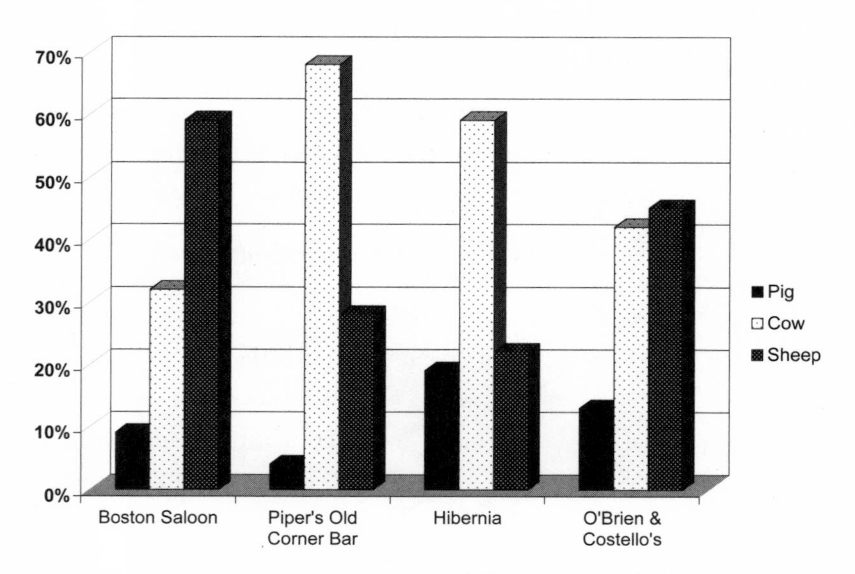

GRAPH 4.2. Percentages of pig, cow, and sheep recovered from the four Virginia City saloons.

each saloon site. Zooarchaeologists, scientists who study animal bones, have found a similar prevalence of domestic animals in the archaeological remains of a rural tavern in Wisconsin, an urban oyster saloon in Sacramento, and a Klondike Gold Rush saloon in Skagway, Alaska.[33] Indeed, the abundance of domestic mammal bones at the four Virginia City saloons, such as those of sheep, cows, and pigs, supports this trend, connecting the Comstock with a national saloon culture. Even though mammal bones from the four saloons appeared to be relatively similar, suggesting that lamb, beef, and pork were popular meat menu items in Virginia City drinking houses in general, the individual excavations revealed subtle differences when the percentages of mammal species served at each place were examined (graphs 4.1 and 4.2).

Domestic mammals prevailed, with negligible amounts of small mammals such as rabbits, as well as fish and fowl, but the bones from the Hibernia Brewery denote that a few wild birds and fish were included on the menu of that establishment. Although small in number in relation to the mammals, such menu items illustrate the Hibernia's relatively low socioeconomic position in relation to its fellow saloons.[34]

The Boston Saloon's high number of sheep remains is another anomaly (see graphs 4.1 and 4.2). Without doubt, lamb constituted the major portion of that establishment's selections.[35] While the numbers are not as extreme as the Boston Saloon's sheep frequency and percentages, O'Brien and Costello's also had sheep as its most frequently occurring meat. Cow bones from Piper's and the Hibernia indicate that beef was perhaps more prevalent on the menu at those places than at the Boston Saloon and O'Brien and Costello's.[36] Pig bones make up the smallest number of these common domestic mammals at all four saloons. The high frequency of lamb, followed closely by cuts of beef, and the small amounts of pork represent a common pattern that emerged in another Comstock-era bone collections from northern Nevada; it may be that sheep could be conveniently slaughtered and their meat used before it spoiled, and that sheep and cows were more available locally than other species were.[37]

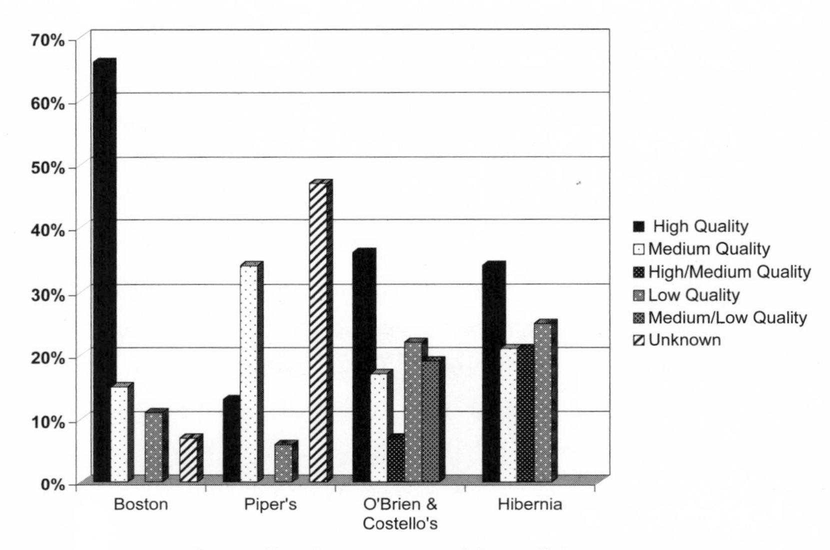

GRAPH 4.3. Qualities of beef cuts recovered from all four Virginia City saloons; the Boston Saloon clearly offered the highest quality of beef.

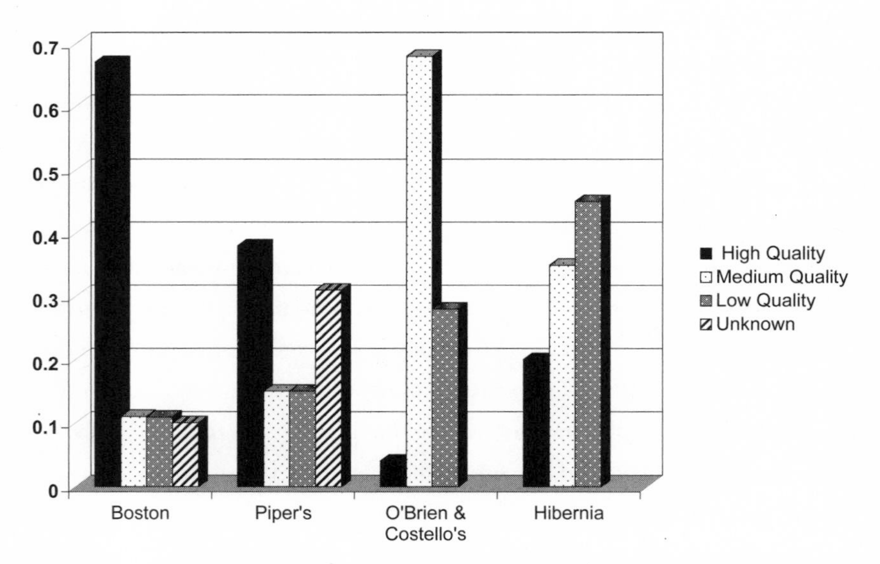

GRAPH 4.4. Qualities of sheep cuts recovered from all four Virginia City saloons; much like beef, the Boston Saloon clearly offered the highest quality of sheep.

While the animal bones give some indication of the types of foods being consumed at the various Virginia City saloons, the body parts they represent tell the story of what cuts of meat were used. Meat cuts were designated high, medium, and low quality depending on the amount or tenderness of the meat. Because some of the saloons offered more high-quality meats than others, the bones help demonstrate additional distinctions between those businesses.

Graphs 4.3 and 4.4 provide an overview of the varying quality of cow and sheep cuts across the four Virginia City saloons. The bulk of the bones from Piper's Old Corner Bar, aside from unidentifiable items, are high- and medium-quality meat cuts. Among these, medium-quality beef cuts were the most significant meat-based meals associated with the theater saloon. The Boston Saloon, on the other hand, had a much larger percentage of high-quality beef cuts, with nearly twice as many as the Hibernia Brewery or O'Brien and Costello's Saloon and Shooting Gallery and nearly three times as many as Piper's.[38] The Boston Saloon continued this trend with sheep bones, illustrating that it served relatively expensive, high-quality cuts of beef and lamb.[39] All in all, the evidence of the bones shows that the Boston Saloon offered the finest meals of all the establishments examined here.

Despite the abundance of expensive beef and lamb at the Boston Saloon, the small number of pig bones from that establishment suggests that low-quality foods were occasionally served. The bones found there include a portion of a skull, some feet, and a few rib fragments.[40] The pig skull is of interest because the location of the saw marks on the teeth (figure 4.13) suggests that the pig head was sawed so that it would sit flat on a surface such as a platter. On the one hand, a pig's head on a platter denotes a festive occasion. On the other hand, pig and sheep heads and feet are often associated with consumers at the lower end of the socioeconomic scale: "In the nineteenth century, public [drinking] houses . . . in humble neighborhoods . . . [offered] such dainties as 'sheep's trotters,' sheep's heads, pig's faces, faggots, etc. . . . hot at certain hours of the day."[41] Whether the pig's head represented a festivity or

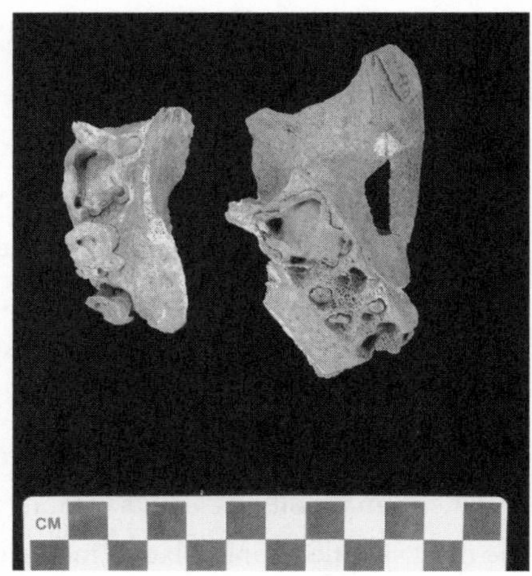

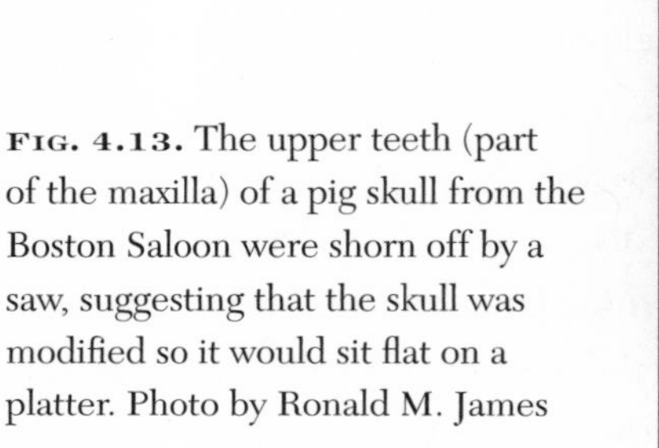

Fig. 4.13. The upper teeth (part of the maxilla) of a pig skull from the Boston Saloon were shorn off by a saw, suggesting that the skull was modified so it would sit flat on a platter. Photo by Ronald M. James

a cheap meal, overall, the Boston Saloon served rather expensive, high-quality, meat-based meals more often than meals using lower-cost meats.

While the Boston Saloon was at the extreme high end of food service among the four saloons, the Hibernia Brewery had the largest amount of low-quality beef and sheep cuts. This, along with the relatively high number of "humbler" foods, suggests that the Hibernia's menu items were cheaper and of lower quality than the fare at the three other drinking houses. The animal bones excavated from the Hibernia provide additional evidence that it was a rather modest establishment.

The Hibernia Brewery and O'Brien and Costello's Saloon and Shooting Gallery appeared to be representative of the lower end of the socioeconomic scale with respect to Virginia City saloons. Even though the animal bone collections from these two saloons initially appeared to be quite similar, with comparable percentages of high- and low-quality cuts of beef, zooarchaeologist Elizabeth Scott noted that slight differences between those collections point to a somewhat higher socioeconomic position for the Shooting Gallery in terms of the types of food consumed.[42] For example, the Hibernia had more medium-quality beef, but an absence of the more valuable calf bones. O'Brien and Costello's, on the other hand, had more calf bones. The Shooting Gallery also had a higher percentage of meaty cuts of pork than did the Hibernia, where the pig bones from the Hibernia included many foot bones. Sheep cuts at O'Brien and Costello's also had a higher percentage of better-quality, meaty bones from the shoulder, rack, and breast areas than did those of its Irish-owned counterpart. Many of the bones from the Hibernia came from pig and sheep feet.[43] Actually, the number of pig foot bones from the Hibernia is more than twice that of O'Brien and Costello's establishment, while the number of sheep foot bones at the Hibernia is almost fifteen times that of O'Brien and Costello's. Elizabeth Scott observed that the presence of foot bones may indicate either secondary butchering at these saloons or the consumption of items such as pickled pig's feet (see graphs 4.3 and 4.4).[44]

Such low-quality meals are not necessarily part of the widespread local memory of Virginia City's boomtown wealth. Rather, local folklore tends to highlight the more opulent saloon menu items, indicating that people from all walks of life enjoyed champagne and oysters, symbols of the boomtown's ostentatious wealth, in the community's drinking houses. The prevalence of such local collective memory became apparent during the excavation of the archaeological remains of places like Piper's Old Corner Bar and the Boston Saloon, as community members frequently stopped by the sites to watch the crew at work brushing the dried clay from various objects, and to ask how many oyster shells and champagne bottles were turning up. While the latter emerged repeatedly, usually in fragments, oyster shells appeared much less frequently. Even though a few showed up at all four saloons, their relative paucity suggests that the folklore has exaggerated the lavish nature of certain menu items in Virginia City's drinking houses (figure 4.14).[45] Specialty businesses were described as "oyster saloons," and it is likely that the existence of such places, along with historical reports of miners dining on various delicacies, encouraged folklore centered on the splendor to be had in the community's saloons.[46] Indeed, archaeological investigations of such places would surely yield piles of oyster shells. Another factor to be considered is that oyster shells are scattered throughout the community; they randomly erode out of the mountain city's sloped embankments and were apparently a nuisance to bottle diggers working in the Virginia City dump in the 1960s.[47] Their prevalence underfoot and in the town dump also explains some of the folklore and reminds us that many Virginia City residents did enjoy such delicacies; unfortunately, the nature of the dump prevents an association of oysters—or any other objects contained therein—with a specific home, business, or saloon.

Whether it was the occasional oyster from the East Coast, high-quality lamb and beef, a low-quality sheep "trotter," or a pig head on a platter, the meat products offered at the various Virginia City saloons illustrate how faunal remains reveal the complexities and distinctions among the foods that people ate while socializing in those establishments. The less-expensive cuts of meat served at the Hibernia Brewery and O'Brien and Costello's Saloon

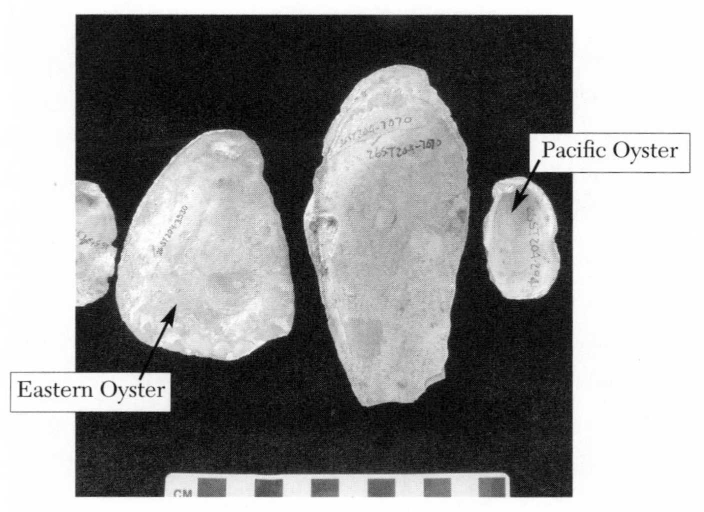

FIG. 4.14. Oyster shells found at Virginia City saloons excavations included the Pacific oyster (*left*) and eastern oyster (*right*) from the Boston Saloon. Photo by Ronald M. James

and Shooting Gallery, in comparison with the more-expensive cuts served at Piper's and the most-expensive cuts at the Boston Saloon, underscore the ways in which Virginia City drinking establishments represented different points on the socioeconomic continuum. Meals are more subtle indicators of a saloon's place along that scale than interior decor is. For example, the most physically and outwardly opulent saloon examined, Piper's Old Corner Bar, did not serve the most costly cuts of meat. It thus appears that these saloons were neither entirely extravagant nor entirely plain on all accounts. They were anything but simple.

A discussion of cuisine would not be complete without considering types of foods that did not survive the passage of time, such as cheese, bread, fruits, and vegetables. At least two peach pits, two apricot pits, and eggshell fragments were found in the dump area of the Boston Saloon, providing evidence of additional and more typically poorly preserved foods that may been associated with this establishment. Such items did not turn up at the other saloons, but it is important to be aware that all drinking houses, if they offered meals, served more than meat-based food.[48]

Condiments

This discussion would also be incomplete without mentioning the condiments that flavored the meals taken at these Virginia City saloons. Archaeology volunteer Dan Urriola mended twenty-one colorless glass fragments to form a small bottle that looked like a pharmaceutical bottle. The base of the bottle, however, revealed the embossed label TABASCO // PEPPER // SAUCE, indicating that its original contents were not medicinal (figure 4.15). This Tabasco bottle, with its thin lip, angular shoulder, and telltale label, turned out to be something of a "missing link" in the pepper sauce company's bottle chronology, becoming the only known example of a transitional form of Tabasco bottles from the company's earliest years of operation (figure 4.16).[49]

In 1868 Edmund McIlhenny produced his first commercial batch of pep-

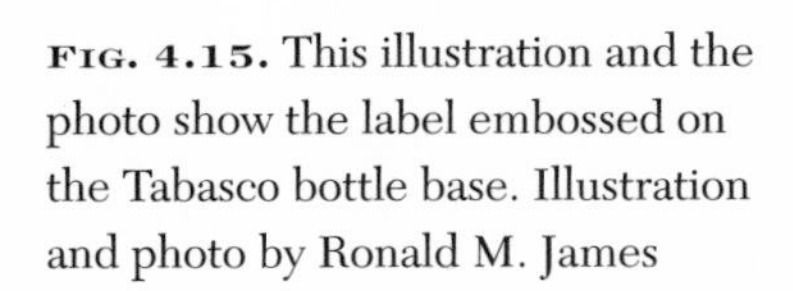

FIG. 4.15. This illustration and the photo show the label embossed on the Tabasco bottle base. Illustration and photo by Ronald M. James

FIG. 4.16. The Tabasco bottle after archaeology volunteer Dan Urriola restored it from fragments. Photo by Ronald M. James

per sauce on Avery Island, Louisiana, using the pepper *Capsicum frutescens* and secondhand cologne bottles.[50] By 1869 McIlhenny made bottles especially for his cleverly named Tabasco sauce; the bottle type at the Boston Saloon may represent one of the earliest of these special-made containers.[51] Because the bottle was found in the lower stratigraphic units of the Boston Saloon site, which are affiliated with the 1875 fire deposits, it represents the temporal window between the date of the fire and the earliest possible manufacture of the specially designed Tabasco bottle, around 1869 or 1870. The pepper sauce company, however, has no records of the product being shipped to Nevada during that period. Because of this, it is easy to imagine that someone carried the bottle to the West. It is also possible that the bottle made its way there via San Francisco shipping routes. Because it was one of the earliest Tabasco bottle types, it represented one of the earliest taste preferences for, or experiments with, this product as a condiment.

It is also significant that this product from Avery Island made its early appearance in the mining West at an African American saloon. Its presence at this establishment and its absence from the other Virginia City saloons may suggest an affiliation with African American cuisine or beverages, given the evidence for pepper sauces in many traditional African, Afro-Caribbean, and African American dishes.[52]

While Tabasco sauce appears to be a rarity in historic archaeological deposits in the mining West, Lea and Perrins Worcestershire sauce is much more common. Lea and Perrins bottles were found at all four saloons and at a residence on the outskirts of Virginia City's Chinatown.[53] On the basis of its consistent presence in the archaeological record of the various Virginia City sites, Lea and Perrins appears to represent a rather common condiment of the period. This product is actually quite universal in historic archaeological assemblages throughout the American West. In fact, its prevalence in this region is said to represent a general means of aiding the discomforts "arising from bad or irregular cooking."[54]

Meal Scavengers

Whether the saloon cuisine was good, bad, or spicy, there were always some things that did not get eaten and that eventually became garbage. People used designated dumps for their refuse, such as the town dump full of oyster shells, but they also tossed garbage into privies and alleyways throughout the city, and the area of the Boston Saloon provides archaeological evidence of this. While digging this site, archaeologists found most of the animal bones and condiment bottles in a dumplike deposit in the alley behind the business. During excavation, the crew unearthed another type of animal bone from this area and from underneath the saloon's charred floorboards, but these bones did not represent meals eaten by people. Rather, they were rat skeletons.[55] Bones from these rodents were quite different from the larger, butchered remains; they lay among the archaeological layers in whole and completely articulated little skeletons. Nobody was butchering and eating rats; the rodents' unmodified bones confirm that fact. Most of the bones were from juvenile rodents, implying the location of a rat nest beneath the floorboards of the Boston Saloon. Obviously, the presence of waste from that establishment's fine meals in the alley dump attracted such scavengers.

The rodents draw attention to the ways in which bones can tell yet another story about urban life in boomtowns such as Virginia City. This city's downtown corridor was composed of many sights, smells, and sounds, and the sight of rats scurrying under and between the dense ranks of buildings provided one of the sensory experiences of that urban setting. Virginia City's magnificence prevailed nevertheless, with saloons that boasted fine food, ornate interior fixtures, and fancy serving ware.

5

A TOAST TO THE ARTIFACTS

Saloon Serving Ware

In television and movie westerns, brawls between the good guys and the bad guys usually featured glass whiskey bottles being smashed over various people's heads[1] or shot glasses being shattered as somebody went sliding down the bar. While some scenes showcased more upscale establishments, with brass fixtures, mirrors, and crystal stemware, many others featured rustically equipped establishments, with sparse sets of utilitarian glassware.[2] This tendency created a bias that undeniably influenced this archaeologist's expectations of saloon artifacts. The depth of this preconception became evident one day when I was wandering through the maze of crew members in the square excavation units as they dug and dusted scatters of artifacts at the Boston Saloon site. Hunched over a pit, one of the field crew members brushed away the dry, gritty clay to expose the glimmer of a crystal wineglass stem.

In the gleam of the northern Nevada sun, that tiny, dazzling object aroused a momentary flash from the past that had the power to unravel the rustic Hollywood renditions. Holding the clay-encrusted crystal stem fragment up in the air, I imagined how it was once part of a complete wineglass that glistened in a dazzling row of crystal champagne flutes and brandy snifters along the back bar of a busy saloon. As I considered the other artifacts recovered from this place, all the bits and pieces began to come together to re-create a fleeting instant from the past. The yellow-tan grit in the air outside the glass windows of the saloon contrasted with its shiny tin ceiling, gaslights, and stemware. Together with the decor, these radiated an ambience of elegance and refinement.

Glassware

As a material remnant of sophistication, crystal stemware calls attention to at least one way in which archaeology can correct the stereotype of the rugged western saloon. Even though the wineglass fragments evoke a counterimage to the sensationalized image of drinking places, such fancy objects were not the primary kind of glassware in use at the four Virginia City establishments. Colorless glass tumblers with faceted bases and glass beer mugs were among the materials shared by all of the saloons, whatever their position on the socioeconomic scale of drinking houses, suggesting that such wares were likely standard saloon equipment (figure 5.1). Similar to other common saloon objects such as green glass wine, champagne, and ale bottles, those tumblers represent products of mass production and mass distribution that thwart interpretations of distinctions among the various saloons.

5.1. Glass tumbler with faceted base, typical of the Virginia City saloons. Photo by Ronald M. James

Fig. 5.2. Beer mug from Piper's Old Corner Bar; a similar object, represented by a beer mug handle exhibiting burn damage, emerged during excavations at the Boston Saloon. Photo by Ronald M. James

Less evenly distributed than glass tumblers, stemware provides a means of understanding at least some of the differences between the establishments. Piper's Old Corner Bar and the Boston Saloon had relatively large amounts of stemware. On the other hand, archaeologists found a paucity of such fancy drinking glasses while digging the other two, less prestigious establishments. By combining historical information about the location and ownership of the four saloons, we could conclude that it is likely that the glasses from which saloon patrons drank reflected the status of each drinking place, with the nicer stemware at places like the Boston Saloon and Piper's Old Corner Bar illustrating the higher status and refined atmospheres at those establishments. The lack of such stemware at the Hibernia Brewery suggests that the Hibernia was the least fancy of the four saloons. O'Brien and Costello's establishment had a few stemware fragments among its repertoire of drinking vessels, and while these did not add up to a mass of fancy wares, they, along with the ornate decanter stoppers, did serve to indicate that this Barbary Coast bar maintained at least some semblance of mate-

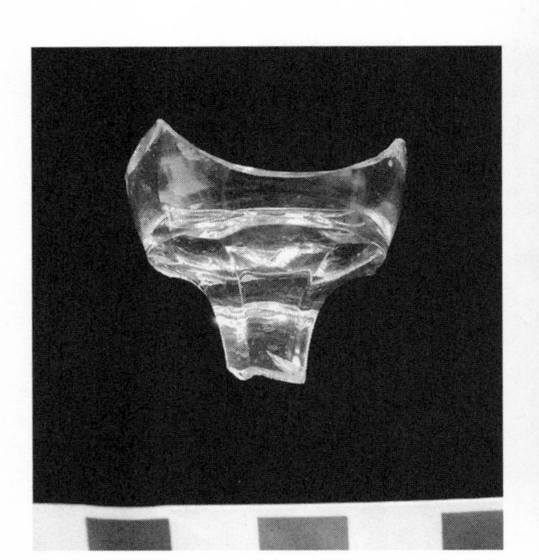

Left: **Fig. 5.3.** One of several pieces of crystal stemware discovered at the Boston Saloon. Photo by Ronald M. James
Right: **Fig. 5.4.** Crystal stemware goblet and stem fragments from the Boston Saloon excavation. Photo by Ronald M. James

rial sophistication. Nevertheless, the higher numbers of stylish pieces at the opera house saloon and the African American saloon imply that those two businesses exhibited more upscale surroundings than the Irish-owned establishments and that they perhaps had a wider selection of beverages.

Glass beer mugs and crystal stemware recovered from Piper's Old Corner Bar and the Boston Saloon portray a diverse array of vessels that served an equally varied assortment of menu items (figure 5.2). For example, the Boston Saloon had a few items indicating that it had a matched set of wineglasses, with smooth, crystal globes and six-sided, faceted stems (figures 5.3 and 5.4). Clearly patrons at this saloon drank wine, and when they did they sipped from the finest glassware. Additional pedestals from crystal stemware indicate the presence of small, delicate liqueur or aperitif-style glasses, suggesting that some customers sipped beverages like sherry or cognac while

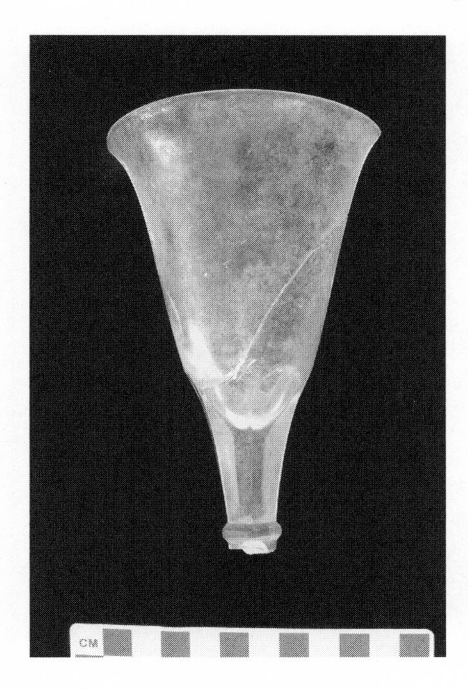

Fig. 5.5. An example of the crystal stemware found during excavations of Piper's Old Corner Bar. Photo by Ronald M. James

socializing at the Boston Saloon; interestingly, the diameters of those delicate pedestals are all different, indicating an even greater number of stemware styles used in that establishment. Piper's Old Corner Bar had its own selection of crystal stemware, including flutes, goblets, snifters, and cordial glasses, which hint at menu items such as champagne, wine, and brandy (figure 5.5).

Ceramic Wares

While glassware distinguished the Virginia City drinking houses, pottery serving vessels shed light on meals at these places. Aside from objects such as the stoneware mineral water jugs and the water filter discussed earlier, most pottery vessels point to food service. Such items included ceramic dinner plates, platters, bowls, and saucers, which were unearthed at all four establishments during the archaeological digs. The quantities of ceramics were still relatively small in comparison to the abundant bottles and other

FIG. 5.6. Stack of ceramic plates at the Boston Saloon, shown partially exposed in the upper left corner of the photo, with a single broken plate to the right of the stack.

beverage containers found at each place. This dichotomy serves as a reminder that even though eating was an important facet of saloon life, it still was not as common as drinking in places dedicated to serving beverages.

Archaeologists discovered the largest number of ceramic dining pieces at the Boston Saloon, which implies that it emphasized meal service to a greater extent than the other drinking places did; this may further explain the evidence that high-quality meat was served there. As the Boston Saloon crew cleared away ash and debris from the 1875 fire, they exposed an entire stack of white ceramic dinner plates sitting on charred floorboards (figure 5.6). Even though the plates had clearly crashed to the floor during the fire, breaking into hundreds of fragments, they had maintained an articulated, stacked formation, as if they had fallen from their storage place on a shelf.

Porcelain knobs and iron hinges lay near the stack of broken plates, furnishing evidence of a cabinet that once held the saloon's dinnerware. The doorknobs were chipped and badly burned, further attesting to damage from the fire, which likely burned the cabinet out from under the plates.

Unlike the saloon's ornate glassware, the plates were plain, made of undecorated white earthenware with a clear glaze (figure 5.7). Although they appeared to be the same style, a closer look revealed that different manufacturer's marks were stamped onto the base of each plate. The marks indicated that the plates came from various potteries in Staffordshire, England, including Powell and Bishop and J. and G. Meakin. Staffordshire potteries mass-produced ceramic vessels during the eighteenth century and by the turn of the nineteenth century were shipping them around the world.[3] The prevalence of these objects at historical archaeological sites is a reminder of

Fig. 5.7. Stack of ceramic plates from the Boston Saloon after cleaning and mending by archaeology volunteer Dan Urriola. Photo by Ronald M. James

a global system that connected outposts the world over with the economy of England and with the expansion of its industries. Even isolated communities like Virginia City, Nevada, were part of this system.[4]

Each pottery company stamped its name or emblem on the underside of its wares. Because mark styles tended to change over time, they have become tools that archaeologists can use to ascribe dates to certain objects. The maker's marks on the plate bases from the Boston Saloon designate a wide span of time, namely, the last fifty years of the late nineteenth century.[5] The various manufacturer's marks also reveal that the business's dinnerware was an assortment of white ceramic plates rather than a matched set.

The cabinet that held the stack of plates also contained other pieces of dinnerware, such as white ceramic serving bowls and platters, decorative porcelain serving dishes, and "yellowware" bowls and platters (figure 5.8). The paste or clay fabric that forms yellowware is a yellow or buff color covered with a clear glaze; *glaze* refers to the glassy coating, usually made from silicate mixtures and bonded to the ceramic surface to provide a protective

FIG. 5.8. Yellowware serving bowl from the Boston Saloon artifact assemblage, after mending by Dan Urriola. Photo by Ronald M. James

and decorative veneer. While yellowware vessels have been available since 1840, their popularity peaked between 1870 and 1900, a period that includes the years when the Boston Saloon was in operation.

In addition to the ceramic service items, crockery made of stoneware was also excavated at the Boston Saloon (figures 5.9 and 5.10). Stoneware is a nonporous, hard, grainy ceramic that ranges from buff to light gray in color and that commonly occurs at archaeological sites as bottles and crockery. Archaeologists dug up shards from a matched set of two large stoneware crocks, one of which was nearly completely reconstructed by Dan Urriola, an archaeology volunteer. Mending also revealed that decorative floral designs covered the entire top of each crock lid. Crockery like this was commonly used for storage. It was likely associated with kitchen storage at the Boston Saloon, a conclusion made possible by using forensic techniques to identify a stain on one of the crock lids.[6] Together with the plates and serving dishes, these vessels imply a greater emphasis on food service and perhaps more frequent dining at the Boston Saloon than at the other establishments.

FIG. 5.9. Stoneware crock and lid with embossed floral decoration, after reconstruction, revealed more of the Boston Saloon's story.

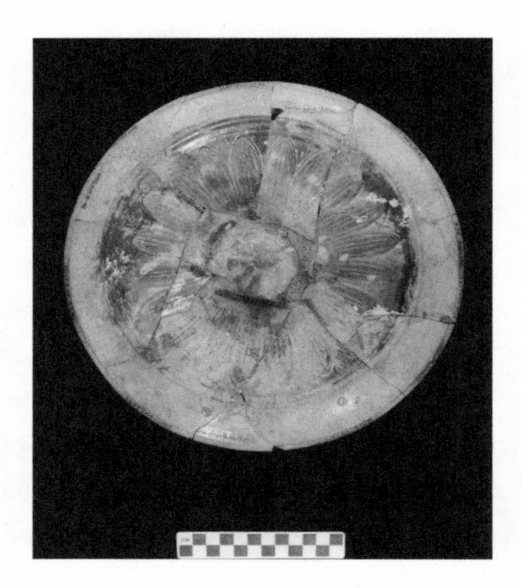

FIG. 5.10. Detail of stoneware crock lid, showing floral design. Photo by Ronald M. James

After devoting so much attention to the Boston Saloon's pottery remains, it is essential to discuss the smaller pottery collections at the other Virginia City saloons. Only a few white ceramic plates, one platter, and one serving bowl came from the excavation units at Piper's Old Corner Bar. These were the same type of undecorated white items as those found at the Boston Saloon, which means that even though both of these businesses offered drinks from fancy glassware, they served meals on rather plain vessels. Archaeologists found similar basic white pottery platters, plates, and soup plates at O'Brien and Costello's Saloon and Shooting Gallery. At least one serving bowl from the Shooting Gallery has a decorative pedestal, signifying that that Barbary Coast establishment had a few ornamental items on hand.

Pottery objects from the three Virginia City saloons noted above were either mended or were sufficiently intact to identify vessel types, such as plates and platters. Unfortunately, archaeology crews recovered no more than ten white pottery shards from the Hibernia Brewery's stratigraphic layers. The pieces were relatively small, but it was still possible to tell that some fragments represented items such as plates or platters. Such a small number of broken pottery fragments may seem to constitute little data about the meal service from this archaeological site, but even this slight amount of information can be useful. For example, a scarcity of such items suggests that they played a relatively minor role in the food service at this saloon. Also, the field crew recovered only a few animal bones, or faunal remains, from the Hibernia.[7] The combination of small amounts of pottery and only a few bones from this site can serve as multiples lines of evidence, indicating that the Hibernia placed less emphasis on food service than did the other saloons.

Another noteworthy fact is that pottery serving vessels from all four Virginia City saloons indicate that they used cheap, undecorated white ceramic. Previous interpretations of archaeological collections from places like O'Brien and Costello's Saloon and Shooting Gallery and the Hibernia Brewery focused on how such utilitarian serving ware reflected the working-class orientation of those establishments.[8] Those interpretations, however, were made before Piper's Old Corner Bar and the Boston Saloon were excavated. Now, with evidence from four saloons instead of just two, it is necessary to reexamine certain objects.

The plain ceramic wares were common in all four of the Virginia City saloons, regardless of their socioeconomic affiliation and their associated extravagant fixtures. What does this mean? To answer this question, it is essential to bear in mind that all four businesses were saloons first and places that served food second. Even though establishments like Piper's Old Corner Bar and the Boston Saloon flaunted their classy atmospheres through novel interior fixtures and sparkling crystal stemware, either they were not as concerned with sustaining airs while serving food to their patrons or they were simply using a standard, sturdy form of restaurant service that could withstand commercial use and reuse. When combined with the other relatively utilitarian fixtures at the two less respectable saloons, the pottery items seem like just another category of austere materials that fostered the plebeian nature of those establishments. However, when put into the bigger picture as a shared material trait across all four saloons, the pottery represents a point of common ground for these places.

Some ceramic serving vessels were more fancily decorated than the plain white plates and platters. These items included teacups and saucers, shedding light on another form of service offered at some of the Virginia City saloons. For example, excavations at Piper's Old Corner Bar uncovered objects like porcelain teacups adorned with gilding and hand-painted floral designs.[9] The Barbary Coast saloon owned by O'Brien and Costello also contained rather fancy pottery associated with tea service, including a white teapot or creamer lid with a floral transfer print and a teapot glazed with a marbled brown "tortoiseshell" design. Among the other items from that site were two small saucers with brown and green floral transfer designs and a porcelain saucer with a blue floral transfer design with gold and black bands. The array of tea service vessels at O'Brien and Costello's appears to be the most diverse among all the saloons examined, a surprising discovery given the nature of a combination saloon and shooting gallery and given the notorious Barbary Coast location. Ironically, this refinement might have resulted from teetotalers' frequenting this Barbary Coast establishment, or it might represent other activities that went on in the same building, such as lodgings on the second floor or a later sewing establishment that shared the first floor with the saloon.[10] Either way, the evidence supports the imagery of the contrasts in Virginia City during its heyday. Whether such settings included shooting gallery participants drinking from fancy teacups or whether they consisted of ladies drinking tea and sewing near the raucous atmosphere of the notorious saloon, Virginia City persisted in its complexities.[11]

Serving Vessels in Virginia City Saloons

Although archaeological research certainly underscores many intricacies, perhaps the most important consideration with serving ware among these saloons is that details of such mundane objects were rarely, if ever, discussed in the historical documents. Advertisements in local newspapers might highlight menu items, but proprietors apparently had no reason to squander promotion space by describing their glassware or dinner plates. Even an advertisement for the sale of a business called the Greyhound Saloon, which was being sold with all of its contents, did not provide details about serving vessels. Everything else was highlighted in the ad, however, with the owner selling "everything necessary to carry out the business," including "beds, bedding, billiard table, bar fixtures, tables, chairs, [and] stoves."[12] Glassware, and perhaps dinnerware, were likely among the Greyhound's "bar fixtures," but the details of these items will be known only if someone re-locates the establishment and carries out an archaeological excavation there. The archaeological research at Piper's Old Corner Bar, the Boston Saloon, O'Brien and Costello's Saloon and Shooting Gallery, and the Hibernia Brewery provided some details on these underdocumented snippets of saloon life in Virginia City.

The discovery of such secrets hidden by the past make it possible to understand moments in the everyday workings of various saloons. Dinner plates and platters may, to archaeologists, seem obviously indicative of meal service, but this has not been part of the more traditional, saloon-brawl images popularized by Hollywood. A brawl in which someone got struck by a plate full of food would indeed provide a vibrant twist to the common story of someone getting hit on the head by a whiskey bottle.

Ideally, this information has nudged the reader's image of saloons away from a scene of brawls and broken bottles and toward a picture of people relaxing while sipping brandy from crystal snifters and eating meals from white ceramic plates. Customers were not necessarily spending their recreation time going to saloons to get into fights. Rather, they frequented the establishments to imbibe, dine, and enjoy a social life. They also engaged in other leisure activities there, such as games, gambling, and smoking, and the archeological evidence of these pursuits provides an even more complex picture of the scene taking place beyond the saloon doors.

6

DESIRES FOR DIVERSION

Saloon Vices and Amusements

Lost in a snowstorm somewhere between Virginia City and the Twenty-Six-Mile Desert, Samuel Clemens and two forlorn companions settled in to embrace their "last night with the living."[1] After the desperate men unsuccessfully tried to ignite a campfire using a revolver shot, their horses bolted off into the snowdrifts, dunes, and darkness. Huddled together in their so-called final hours—without horses and without a campfire—the men awaited the "warning drowsiness that precedes death by freezing."[2] At that point each of them pledged to reform himself of certain vices if they should miraculously live through the night. One man swore that he would quit drinking whiskey. The second pledged to give up his card-playing fetish. The third, Clemens, vowed to throw away his tobacco pipe. Inexplicably, the three companions survived the night and, when daylight arrived, discovered that they had been stranded only a few steps from a warm inn, where their horses had taken shelter. Needless to say, all three soon abandoned their vows to reform and returned to their respective vices.

When the archaeological evidence is merged with the vices described in Mark Twain's winter storm story, it becomes clear that drinking, smoking, and gambling represented a trinity of habits that were quite common in boomtowns.[3] With the recovery of artifacts such as tobacco pipes, dice, and poker chips alongside the array of beverage containers and glass serving ware, archaeological projects in Virginia City revealed that saloons provided havens for the expression and enjoyment of those three everyday activities that were viewed as vices or indulgences. These objects also help to make contact with the interior atmospheres of each business, revealing the ways in which alcohol, tobacco, and various forms of gambling and game playing permeated each place.

Tobacco Paraphernalia

Tobacco pipe bowls and pipe stem fragments surfaced as archaeologists dug through scatters of artifacts lying about the ruins of each saloon site (figure 6.1). These objects indicate that people smoked tobacco at each public drinking house and that tobacco use accompanied social drinking in boomtown saloons. A comparison of pipe fragments from the four sites made it clear that plain white clay tobacco pipes were the most prevalent tobacco-

Fig. 6.1. White clay tobacco pipe fragments, the most common type discovered during Virginia City saloon excavations. Photo by Ronald M. James

smoking paraphernalia at these saloons; such pipes are common finds across nineteenth-century archaeological sites in general. In this respect all four establishments were alike, despite the fact that their patrons likely represented various ethnicities and social classes. The saloons expressed differences, however, in terms of their respective quantities of these objects, with the Boston Saloon yielding the highest frequency of plain white clay tobacco pipes, as well as the highest overall frequency of tobacco paraphernalia.

Most of the white clay pipe bowls and stem fragments from the various saloons are undecorated, but some display manufacturer's marks on their otherwise plain surfaces. Marks on some of the pipes recovered from Piper's Old Corner Bar and from the Boston Saloon include the letters "T.D." on the pipe bowl, facing the smoker (figure 6.2). The "T.D." mark is characteristic of pipes made by McDougall of Glasgow.[4] A white clay pipe stem fragment that

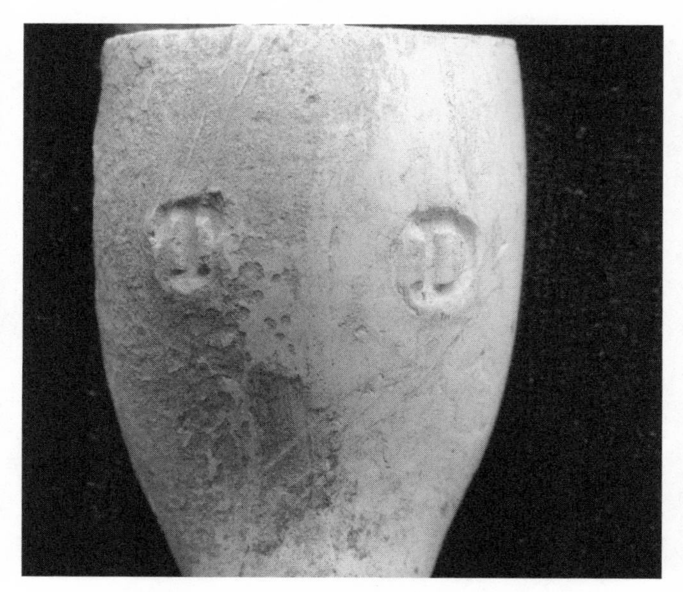

FIG. 6.2. Close-up view of the plain, white clay tobacco pipe showing the "T.D." mark. Photo by Ronald M. James

mends to one of the bowls marked with "T.D." actually contains the words "MCDOUGALL" and "SCOTLAND" stamped along either side of the stem.

A plain white pipe bowl with the "T.D." mark from the Boston Saloon tells its own story. This object is blackened on the inside upper [smoker's] left rim of the bowl. Such charred markings in this area indicate habitual lighting on the smoker's left side of the pipe bowl, which presents an even more vivid image of the Boston Saloon: Someone, perhaps an African American gentleman, sat or stood with his head cocked slightly to one side, holding a match in his right hand to light the tobacco packed into the pipe bowl held in his left hand. Profuse puffs of smoke billowed around his head as the pipe ignited and as he shook his wrist to extinguish the match. Habitual lighting of this pipe in this fashion resulted in the char damage on the left exterior portion of the pipe bowl.

While white clay pipes like those described above were quite common in nineteenth-century artifact assemblages, the Boston Saloon dig recovered another, rather anomalous style of tobacco pipe. This style is represented by at least three undecorated J-shaped red clay pipe bowls (figure 6.3). Each contains a bore hole for inserting a detachable pipe stem into the bowl. Unfortunately, archaeologists did not find any pipe stems that appeared to fit the diameter of the bore holes. This rather plain but rare pipe did not appear in any of the other Virginia City saloons, which suggests that it was a unique style used only by someone who patronized or who worked at the Boston Saloon. Because this type of pipe is unknown to other archaeologists and tobacco pipe specialists,[5] its origins and how it came to be at this African American business in Virginia City, Nevada, remain mysteries.

Excavations at O'Brien and Costello's Saloon and Shooting Gallery yielded at least one red clay pipe bowl, which initially appeared to be similar to the red clay styles recovered from the Boston Saloon (figure 6.4). However, the pipe bowl from the Shooting Gallery is slightly more decorative, displaying two parallel relief bands that encircle the uppermost portion of the curved pipe bowl. The red clay pipe bowls at the Boston Saloon had no

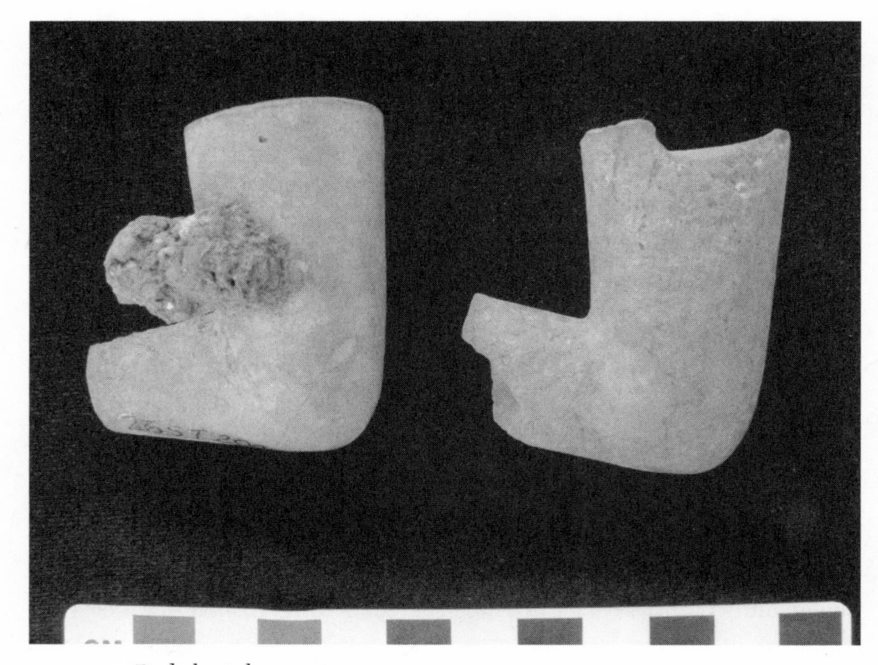

Fig. 6.3. Red clay tobacco pipe bowls, like those shown here, were found at the Boston Saloon; the bowl on the left has a piece of corroded metal fused to its face. Photo by Ronald M. James

Fig. 6.4. Red clay pipe bowl fragment from O'Brien and Costello's Saloon and Shooting Gallery, with two raised bands along its upper rim. Photo by Ronald M. James

such decoration and were more angular in design than the rounded bowl found at the Barbary Coast establishment.

The most elaborate tobacco pipe discovered was excavated at Piper's Old Corner Bar (figures 6.5 and 6.6). This meerschaum pipe bowl emerged from the debris that collapsed into the cellar of the "theatre saloon" after a fire damaged the Piper's building in 1883. Meerschaum pipes are made of a white, claylike mineral that darkens every time it is smoked. The Old Corner Bar pipe bowl had been smoked enough that it had turned a dark brown color. Carved on its face is a relief design featuring three figures: a human dressed in hunting garb, a mythical creature with a doglike face and a human body dressed in hunting attire, and a dog. Both the human and the mythical figure wear ammunition belts and pouches slung across their torsos, and bear rifles. This pipe is carved so intricately that the particulars of the hunting accessories are quite clear, and it is even possible to see the detail of hairs on the dog's body. As the mud and clay were cleaned from this conversation piece during the excavations at Piper's, the archaeology crew began contemplating whether John Piper may have smoked the pipe. The owner may never be known, but the presence of this object certainly adds to the sense of being in this saloon. Whether John Piper stood behind the bar puffing on the sizable pipe or whether another patron enjoyed the pipe while sipping drinks, it was an object that would have been noticed by people in this establishment.

Among the other forms of tobacco paraphernalia that emerged during archaeological excavations was a pipe mouthpiece made of amber, from the Hibernia Brewery (figure 6.7). Teeth marks were evident on one end of this malleable stem piece, while on the other end the mouthpiece was threaded on the inside of the center hole, allowing for its attachment to a pipe bowl. A second threaded, tube-shaped mouthpiece made of bone also appeared during investigations at the Hibernia.

Indulgence in tobacco did not always come in the form of smoking, as shown by a set of pottery spittoons that archaeologists pulled from the

FIG. 6.5. Front view of meerschaum pipe bowl from Piper's Old Corner Bar. Photo by Ronald M. James

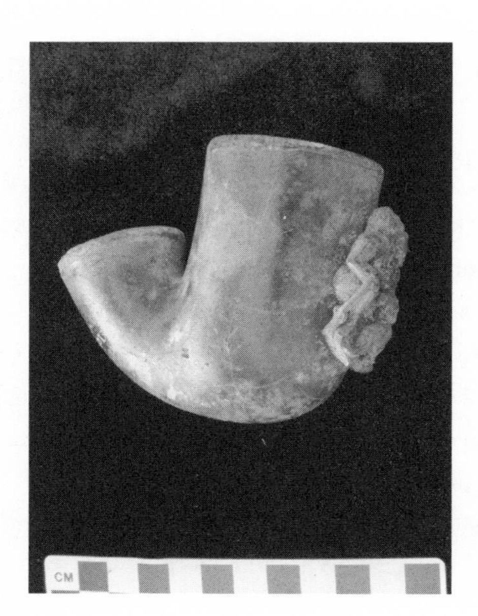

FIG. 6.6. Side view of meerschaum pipe bowl from Piper's Old Corner Bar. Photo by Ronald M. James

Fig. 6.7. Amber pipe stem mouthpiece with bore hole fixture and teeth clench marks, as recovered from Hibernia Brewery. Photo by Ronald M. James

Fig. 6.8. The largest of three matched stoneware Bennington-like spittoons discovered at Piper's Old Corner Bar. Photo by Ronald M. James

depths of Piper's Old Corner Bar. While they provide evidence that people chewed tobacco as well as smoked it, the spittoons also represent some of the fancier fixtures that are unique to the opera house saloon. They have a shiny tortoiseshell glaze, which is represented by a brown-and-tan marbled design and was sometimes noted as "Bennington ware" (figure 6.8). Cracks in fragments of the spittoons revealed that the pottery was made of a buff-colored stoneware. While the three spittoons were exact replicas of each other, they were three different sizes: small, medium, and large, with the largest having nearly a one-foot diameter. The face of each was cast with a stylized banner containing a circular shape in its center and the bust of a bearded man wearing a turban inside the circle.

Gambling and Game Playing

Smoking and other forms of tobacco use are certainly popular, almost excusable, indulgences according to modern standards, but they were a "hatable vice" for some during the nineteenth century.[6] Gambling was and continues to be another fairly widespread addiction. During the nineteenth century and up to the present, Virginia City saloons have consistently hosted gambling activities of various sorts. As one enters any of the refurbished historic saloons in Virginia City today, such as the Bucket of Blood or the Delta, rows of slot machines beep, blink, and entice patrons to try their luck. Together with nineteenth-century writings, which describe poker or faro games, billiards, dice, and card playing,[7] artifacts such as poker chips and dice drive home the point that gambling and game playing were universal, auxiliary vices to drinking and smoking in those businesses. For example, excavations at Piper's Old Corner Bar and the Hibernia Brewery recovered dice made of bone, indicating the presence of dice-related gambling and games in those establishments, and archaeologists at the Boston Saloon recovered fragments of red and grayish-blue poker chips (figures 6.9 and 6.10).[8]

Other artifacts may or may not have been used to satisfy gambling habits.

FIG. 6.9. Dice made of bone from the Hibernia Brewery. Photo by Ronald M. James

FIG. 6.10. Poker chip fragments found during excavation of the stratigraphic layers atop the Boston Saloon. Photo by Ronald M. James

For example, while digging in Piper's Old Corner Bar, the archaeology crew came across a cribbage board (figure 6.11). Cribbage appeared early in the seventeenth century and subsequently became a favorite pub and gambling game among various gentlemen throughout Europe,[9] eventually making its way to saloons in the American West. The game relied on playing cards, and the board provided a means of keeping score by inserting pegs. The board at Piper's was found basically intact, with only a few nicks marring its surface; it was carved from a soft volcanic rock into the shape of a small, solid rectangular block. The board's playing face had two lines of sixty-two holes for the playing pegs. Each side of the board displayed decorative carved patterns of various playing card suits, including an upside-down spade, a heart, a diamond, and an upside-down club. People also played cribbage at the Hibernia Brewery, as shown by cribbage pegs collected from that site during archaeological investigations.[10] The evidence of cribbage there and at Piper's suggests that the European and Irish immigrants and the European and Irish Americans in Virginia City enjoyed the game of cribbage regardless of the class status of their respective drinking houses.

Dominoes were another of the games played in Virginia City saloons. A

FIG. 6.11. Intact cribbage board from Piper's Old Corner Bar. Photo by Ronald M. James

Fig. 6.12. This tiny burned and slightly calcined domino fragment made of bone from the Boston Saloon. Photo by Ronald M. James

fragmented domino made of bone, mistaken at first for a burned bit of mammal bone, emerged from the Boston Saloon excavation (figure 6.12). This object exhibited a bluish-gray color, a condition known as "calcining," which was the result of severe burning of the domino's bone material. Dominoes provided an avenue for socializing in the Boston Saloon, as elsewhere: "Banter is devoted to the game, and the game is devoted to the fine skill of socializing."[11]

Objects such as the cribbage board and dominoes certainly could have been associated with gambling. On the other hand, such objects may also reflect simple game playing, yet another form of amusement that was part of the leisure atmosphere in the various saloons. All sorts of games, including poker, cribbage, dominoes, and marksmanship (at a place like O'Brien and Costello's Saloon and Shooting Gallery) went hand in hand with the socializing in Virginia City's public drinking places. The presence of these items also provides a direct link to a typical scene in these establishments, drawing attention to moments during which one would have heard cards being shuffled, the clatter of dice rolling, the slap of dominoes, and the ring of poker chips on tables or bar countertops.

The Mysterious Presence of Women

While saloons were certainly places for people to gamble, smoke, drink, and socialize, they offered other forms of recreation as well. Technically, sex connected with prostitution may also be treated as a vice associated with boomtowns and saloons.[12] Not all places offered such intimate degrees of female company, however; some saloon entertainment included less cozy camaraderie, such as female dance partners.[13] There is no historical evidence of female dancers or prostitutes "working" at Piper's Old Corner Bar, O'Brien and Costello's Saloon and Shooting Gallery, or the Hibernia Brewery. However, artifacts indicative of women at these establishments—clothing accoutrements such as buttons, beads, and other fasteners—were excavated.

Most historical archaeological assemblages include a hodgepodge of mismatched buttons denoting various fasteners lost over time. In many cases these fasteners can be associated with certain types of garments, and they can in turn reflect gender-based clothing styles of the time. In the Virginia City saloons, for example, a few objects from O'Brien and Costello's Saloon and Shooting Gallery may represent women's clothing (figure 6.13). One of these, a black glass button with a conical shape and a series of faceted "beads" forming a design on its face, likely fastened a coat or other relatively heavy garment. Another black glass button was representative of a fastener used to secure women's shoes or boots; this item was small and beadlike, with diamond-shaped facets decorating its surface.

On nineteenth-century sites elsewhere, such as the Boott Mills of Lowell, Massachusetts, black glass buttons have been interpreted as affordable examples of women's efforts to keep up with fashions. Clothing accoutrements made from black glass were substitutes for the more costly "jet" material, which was made from dense black coal. Queen Victoria popularized jet in the 1860s while mourning for Albert, her husband.[14]

FIG. 6.13. Black glass buttons from O'Brien and Costello's Saloon and Shooting Gallery, including the black glass women's boot or shoe button (*right*) and the design with faceted beads. Photo by Ronald M. James

The presence of accessories from feminine clothing implies that women were connected with this leisure establishment. Given the reputation, or stigma, of women on the Barbary Coast, their presence at the Shooting Gallery may imply that establishment's association with prostitution—or it may simply indicate that the saloon employed women to assist with the bar activities.[15] The rather chic nature of the clothing fasteners found at this Barbary Coast saloon may hold the key to understanding the role of some women there.

Usually women who worked in the liquor trade did so because they needed to help support their families; such women did not likely have the means to dress in relatively fancy apparel. On the other hand, women working as prostitutes probably "dressed up" a bit more than the women working as bar servers. In other words, the women associated with these buttons dressed well, suggesting a provocative presentation. Such attire may not have been all too different from the typical—and more morally respectable—Virginia City woman's apparel; as one outspoken female resident, Mary McNair Mathews, observed, the people of that community were better dressed than any other place she had lived.[16] In reference to prostitutes, however, Mathews goes on to say that the "unmarried" women were always dressed the "richest."[17] Considered in the context of Mathews's remarks, audacious clothing may have reinforced the stigma of the surrounding Barbary Coast, demonstrating how, on occasion, "the line between brothel and saloon often blurred" in that district.[18]

Eventually, outrageous behavior in the Barbary Coast, such as child prostitution,[19] sparked a community effort to reform and rehabilitate that vice-ridden district. The Hibernia Brewery operated on the outskirts of the Barbary Coast around 1880, after the community's attempts to clean up that area had been set in motion. This undertaking complicates interpretations about the presence of women at that Irish-owned establishment. For example, a brass garter buckle surfaced during excavations at the Hibernia (figure 6.14). While it is important to bear in mind that men also used garters, this object could just as likely have been worn by a woman and may imply the presence of women with this establishment. However, the Hibernia was adjacent to a boardinghouse and a Singer Sewing Machine sales office,[20] so it is entirely possible that women's clothing accoutrements associated with those activities could have become mixed with the saloon's archaeological deposits.

Since Piper's Old Corner Bar held the position of a respectable "theatre saloon" in the center of town, fancy clothing fasteners from that saloon could be interpreted as belonging to well-dressed women attending the opera house. Archaeologists expected to find a myriad of such accoutrements, but they did not. One button, with a pockmarked floral design, may have fastened a woman's coat; this object appeared to be rubber but was actually gutta-percha, a substance that resembles rubber but is harvested from the latex of Malaysian trees. It was not as impressive, however, as the women's clothing fasteners found at the notorious Barbary Coast saloon. Neither was the second item from the Piper's collection, a porcelain button. Displaying a brown transfer design on its face, this fastener was known as a "calico button," because it was designed to match calico dress fabric. Such buttons were typically used on women's dresses, particularly on the everyday clothing worn by "lower class" women.[21] Its provenience in the Piper's assemblage suggests the presence of women, and quite possibly women who did not always dress in an upscale, fashionable manner. This could be a result of the fact that Piper's Old Corner Bar, by catering to anyone attending the theater upstairs, likely served the full range of Virginia City theatergoers, some of whom could not afford fancy attire.

Aside from the single calico button, the bigger story to be found in the clothing accoutrements at Piper's Old Corner Bar is the overall lack of artifacts associated with women. This may be because Piper's was, for the most part, a gentlemen's club. According to one nineteenth-century writer, John Piper's Old Corner Bar offered "gentlemen" fine brandies, wines, and cigars; as a matter of fact, gentlemen were tempted almost "against their will" to go out and visit the establishment each evening.[22] That this reference specifies "gentlemen" suggests that relatively respectable members of the

FIG. 6.14. Brass garter clasps from the Hibernia Brewery site. Photo by Ronald M. James

community patronized Piper's saloon. It also suggests, along with the sparse archaeological evidence of femininity, an absence of women with either respectable or disreputable reputations. In other words, women, whether prostitutes or not, generally did not go there.

Something altogether different was going on at the Boston Saloon with regard to women, leisure, and vice. Unlike the other three saloons, about which the historical record remains silent with respect to the presence of women behind their doors, the Boston Saloon, according to one historical record—the Nevada State Census from 1875—had a "courtesan" living on or near its premises. Born in Massachusetts, this woman was named J. Lind. The census enumerator's use of *courtesan* in reference to J. Lind could mean a number of things. It was not unheard of for census enumerators in Virginia City to exhibit prejudices against Chinese women, identifying them as prostitutes even when they were not.[23] Given this propensity to misrepresent Asian women, it is entirely possible that they treated African American women in the same manner, especially an African American woman associated with a saloon in the D Street red-light district, described by some as the "condemned part of the city."[24] It is possible that J. Lind was someone who merely worked in or resided near the saloon. She may also have been William Brown's wife but was not given a valid description by the census enumerator. We know that Brown was married to someone through common law in 1864, but unfortunately the name of his wife and the duration of their marriage have not been found in the historical documents.[25] Alternatively, J. Lind may have been Brown's mistress or a prostitute. There is no way to verify any of these possibilities or to unravel the mystery of J. Lind.

Then again, artifacts such as women's clothing fasteners recovered during excavations of the Boston Saloon do add more to the story (figure 6.15). Shining against the gritty tan clay that hid it for many years, a black glass button emerged from the long-buried African American business. The button reflected iridescent violet and blue tones in the sunlight, conjuring diamond-in-the-rough imagery. Seven faceted "beadlike" pieces were arranged in the shape of a flower on the button's glass backing. Such a button once fastened a stylish woman's coat. A second fastener associated with either a woman's coat or a similar heavy garment also emerged from the excavation pits at the Boston Saloon. This was a square black glass button with beveled edges. This type of button and its associated garments were relatively inexpensive, especially when compared with the fancier iridescent style, and could be described as indicative of a moderately priced garment.[26]

Still more dainty clothing accoutrements appeared as the crew dug on, some of them so tiny that they slipped through the archaeological screens. A series of decorative beads at first appeared to be pebbles encased in crusts of clay. Some of these were made of glass, but their original golden color had long since disappeared. Instead, a glittering tint, or patina, caused by burn damage coated the surface of each bead. These minuscule objects appeared during excavations on the charred floorboards from the Great Fire of 1875. It is quite likely that that blaze was responsible for the modifications to the beads. It is possible that they were used as dress beads. The person who wore

Fig. 6.15. Women's clothing accoutrements from the Boston Saloon. From left: black glass button with relief floral design; square black glass button with beveled edges; white porcelain shank-style or shoe/boot button; fancy black glass button with iridescent glaze. Photo by Ronald M. James

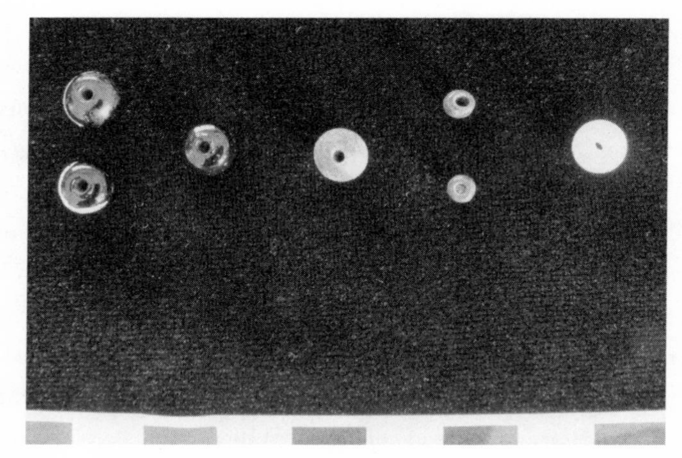

FIG. 6.16. Dress beads from the Boston Saloon. *From left:* black glass beads; cobalt blue bead; samples of burned taupe- or golden-colored glass beads; white glass bead. Photo by Ronald M. James

them, or the beaded garment that they decorated, certainly made a dramatic addition to the ambience of the Boston Saloon.

Other dress beads popped up during excavations at this site, teasing the archaeology crews with hints of the striking gowns worn by women in this saloon (figure 6.16). These included black glass, white glass, and cobalt-blue glass beads, as well as a six-sided tube-shaped bead. Two black glass rings, or hoops, were another type of beadlike object associated with women's attire at this establishment; these rings provided the structure for a dressmaker to sew around so that she or he could use them as templates to create handmade buttons for a woman's dress or outfit.

In light of the relatively small amount of women's clothing fasteners found at the other three Virginia City saloons, the quantity, diversity, and vividness of these objects at the Boston Saloon revealed a major distinction that set this place apart from the others: women—rather well-dressed women—either patronized or worked in this establishment to a much greater degree than they did at the other places. Like many leads that emerge while carrying out historical and archaeological research, these details from the past provided tantalizing evidence but did not actually answer any of the questions they raised. Instead, they created a sort of Cinderella story with regard to the person or persons who donned such costumes. Who was she? Who were they?

Speculation using the snippets of information available in the historical records is, unfortunately, the only means of addressing these questions. Perhaps some of the items were worn by J. Lind, the African American woman described as a courtesan in the 1875 Nevada State Census. Maybe some of the items belonged to saloon proprietor William A. G. Brown's wife, a woman who is mentioned only once in the documentary record and whose name remains unknown.

Another candidate who certainly dressed stylishly was African American entrepreneur Amanda Payne. Payne owned her own boardinghouse and restaurant in Virginia City and also was in business as a saloon owner on D Street for a short time.[27] It also appears that William Brown at some point was working either for or with her in the saloon business. Amanda Payne, clearly a respected and successful Virginia City entrepreneur with the means to dress well, likely participated to some degree in the activities of the Boston Saloon, for business purposes if for no other reason.[28] Although it is not possible to state unequivocally that the clothing accoutrements are directly associated with her, it is possible to state that she could have been the owner of dresses that were graced with such accessories. Payne was, after all, an outstanding citizen in Virginia City as a whole and was also one of the people associated with this establishment.

When examining an array of artifacts such as these linked with women's clothing, it is important to consider as many sources as possible, from people like Amanda Payne to William Brown's wife. It is also necessary to remember that D Street was the location of Virginia City's red-light district. Although most prostitutes worked directly from cribs or brothels, they were certainly visible among the various drinking houses on D Street. In other words, the Boston Saloon's location in a designated vice district could be one factor in an explanation for the extraordinary number and variety of women's clothing fasteners found at that excavation site. Fancily clad women were part of the D Street landscape. The presence of such artifacts in archaeological investigations of that area merely validated the historical uses of the area. Even so, given the sociohistorical context of racism, fancily clad white women were probably an unlikely sight in an African American place of recreation like the Boston Saloon.[29] It is therefore quite likely that the beads and buttons came from dresses worn by African American women.

Toys

Vices, indulgences, game playing, and various other forms of recreation in public drinking houses come to life through a random mix of artifacts. Some of these portray well-dressed women, while others reveal adults at play, sometimes with toys usually associated with children. Among the toys recovered from the saloons were marbles, and because adults used, "played with," and even gambled with marbles in these saloons, any assumptions about those items and children should be made with caution. Similarly, it may seem that porcelain dolls would indicate child's play, but that interpretation, too, must be made with caution, especially in a saloon on Virginia City's Barbary Coast with its nearby brothels and cribs. According to oral tradition in places such as Skagway, Alaska, porcelain dolls served rather utilitarian purposes in businesses that offered female prostitution. Prostitutes who worked upstairs in a saloon kept a line of dolls along a piano in the bar area. Each doll represented one of the prostitutes. If a woman's doll was lying down, that was a sign that she was occupied. If the doll was sitting up, that meant she was available.[30]

A variety of toys appeared during excavations at most of the Virginia City saloons. For example, a handful of marbles came from the Boston Saloon.

Other marbles, a porcelain doll arm, and a tiny porcelain "tea party" cup and saucer were found in an alley on the north side of O'Brien and Costello's establishment. The location of these toys in the alley may support the argument that they were used by children rather than in pursuits like gambling, prostitution, or other vice-laden activities within the saloon. Excavations at the Hibernia Brewery also recovered porcelain doll fragments and marbles. And more marbles and a toy gun came from Piper's Old Corner Bar. While these items cannot unequivocally be associated with children's play, there is one passage in the historical record that mentions some boys underfoot at Piper's Opera House.

According to that account, the boys were always trying to get into the opera house to see some of its shows even though they did not have any money: "Each performance in Piper's Opera House meant forty or fifty penniless small boys hoping for miracles as they stood outside the entrance. Old John Piper would sit in the box office and scowl at us until his face was all wrinkles. Then [John] Mackay would come along, nod his head toward the gang, and say, 'John, how much for the bunch?'"[31] This passage recorded a real-life scenario of children, or at least young boys, loitering around the opera house, which likely meant that they also lingered around Piper's Old Corner Bar before and after attending performances. The archaeological excavations validated the representation of the children as boys, since they did not recover any toys stereotypically associated with girls, such as dolls and tea sets.

Leisure and Vice: Concluding Comments

The line between play and vice becomes blurred with adulthood. Indeed, one person's vice may be another's form of play and relaxation. On the other hand, it is easy to see how alcohol, smoking, gambling, and prostitution were considered immoral habits. Historical writing, such as Mark Twain's story of desperation and vice in the desert snowstorm, combine with archaeological evidence to underscore the ways that stereotypically "bad" habits actually served a purpose, that of providing comfort and recreation. Consequently, artifacts that can be linked with vice and indulgence in saloons can also be reminders that people patronized those places to escape their daily troubles, to relax, and to enjoy a drink, a smoke, and perhaps a game of cribbage.

These activities were interconnected in saloons because such establishments were in the business of providing a venue where people could partake of those forms of leisure. One nineteenth-century writer described a rather seedy scene of vice, noting that "little stacks of gold and silver fringed the monte tables and glittered beneath the swinging lamps . . . the rattle of dice, coin, balls, and spinning-markers, flapping of greasy cards and chorus of calls and interjections went on day and night, while clouds of tobacco smoke filled the air and blackened the roof-timbers."[32]

Admittedly, vices did then and still do play a major role in many people's choices of recreation. This realization, along with the various artifacts from the Virginia City drinking houses, has the potential to overturn the stereotype of saloons as hotbeds of violence and shoot-outs. By catering to the citizenry's desires for rest, diversion, entertainment, and female companionship, these establishments can instead be viewed as places that provided boomtown residents with a valuable service—that is, a chance to escape the isolation and stress of life in a mining camp.

With this understanding, it is important now to revisit Hollywood depictions of western saloons and reflect upon how vices such as drinking and gambling have been portrayed. Barroom brawls have certainly been the cornerstone of action in Hollywood saloon representations, and many of these conflicts result from gambling disagreements.[33] This evidence could leave the impression that the vice of gambling was, for the most part, a dangerous—even deadly—form of recreation. The evidence from the Virginia City saloons, however, illustrates that such activities were not necessarily undertaken with a brawl constantly lurking in the background. Rather, the array of "entertainment artifacts" from the four saloons and the lack of historical accounts of constant brawling in those drinking houses illustrate that gambling, smoking, and other vices were actually activities in which people engaged when they were at leisure, relaxing in various establishments of public drinking.

CRIME SCENE INVESTIGATION?

Forensic Applications and Saloon Artifacts

While artifacts such as tobacco pipes help to re-create the ambience of saloons, a closer, "forensic" look at some of these objects has the potential to expand an understanding of other activities taking place in and around boomtown saloons.[1] Although archaeology shares some methods with crime scene investigations, it deals with much older evidence.[2] Archaeologists lay out grids to map artifact locations, and they treat artifacts like evidence. After assembling clues based on the artifacts and their placement within the layers of a site, archaeologists try to recognize what happened at a particular place. This process is much like a criminal investigator's method of piecing together the events associated with a crime scene.

Forensic science is another discipline that intersects with archaeology. The modern meaning of the term *forensic* expands upon its Latin root, meaning a forum or public place. Today the term refers to public, judicial courts. Forensic science is therefore used to address questions posed by a legal system, and the techniques are expected to stand up to legal debate. Using tools such as fingerprinting and voice printing, ballistics, bone and soft tissue identification, toxicology, and DNA testing, forensic investigators lift microscopic bits of evidence from victims, suspects, and the material scenes of crimes. If preservation conditions are favorable, these types of evidence may survive at much older, archaeological scenes of past events.

Archaeologists working at the Boston Saloon decided to subject some of their artifactual evidence to forensic scrutiny. While digging through the layers of clay at the ruins of that site, they came upon intact but charred floorboards from the building that had once held the saloon. As they brushed away the grayish-white ash and debris to expose untouched saloon materials, the field crew felt as if they had unearthed a crime scene. The excavators took care to leave all the artifacts in place so that they could photograph and map the layer of culture that contained the saloon's last moments, and the crew eventually realized that some materials from the Boston Saloon could be subjected to forensic testing. These tests revealed subtle details about moments in and around that establishment during Virginia City's heyday.

Burn Damage on Bones

Chapter 4 examined ways to understand meal service in the various saloons by using animal bone fragments. A forensic analysis of how bones burn provides additional insight. Thirty percent of the bone sample analyzed had burn damage, with 16 percent of those exhibiting a condition known among forensic anthropologists and zooarchaeologists as "calcining." When bones become charred, the collagen is carbonized, making them appear black. Calcining is the next stage of burn damage after a bone is charred. This condition appears when all organic material, or collagen, has been completely burned out of the bone, leaving the bone in a form that is pure mineral and resulting in colors that range from bluish-gray to white. White is a definite indicator of a bone's calcined condition. The presence of calcined bones indicates that burn temperatures reached 645° C/1,200° F and lasted longer than a typical structural fire. Usually such conditions are caused by incineration.

This evidence suggests that the burn damage on the Boston Saloon's calcined bones may not necessarily have resulted from a structural fire, such as the one that was part of Virginia City's Great Fire of 1875. By stepping away from the bones and returning to the archaeological "scene," it was possible to consider the potential causes of such high temperatures. The faunal remains were primarily recovered from two general areas in the archaeological grid system used at the Boston Saloon. One area was associated with the saloon building, the scene of the major structural fire caused by the 1875 blaze. The second area was an alleyway behind the saloon that contained a dump (figure 7.1). More bones in the dump area had burn damage than did those from the saloon area. Bones in the dump area also exhibited much more extreme burning, containing more calcined specimens, than the bones associated with the Boston Saloon building (graph 7.1).[3]

The evidence exhibited by bones from the dump area suggests that they were intentionally burned over a long period of time and at higher temperatures than normal for cooking purposes or a structural fire. In other words, someone was probably incinerating garbage in the alley dump behind the Boston Saloon. Most bones recovered from this site came from the dump, and most were unburned. The unburned specimens were clearly bones that did not get incinerated in the alley dump area. But more important is the fact

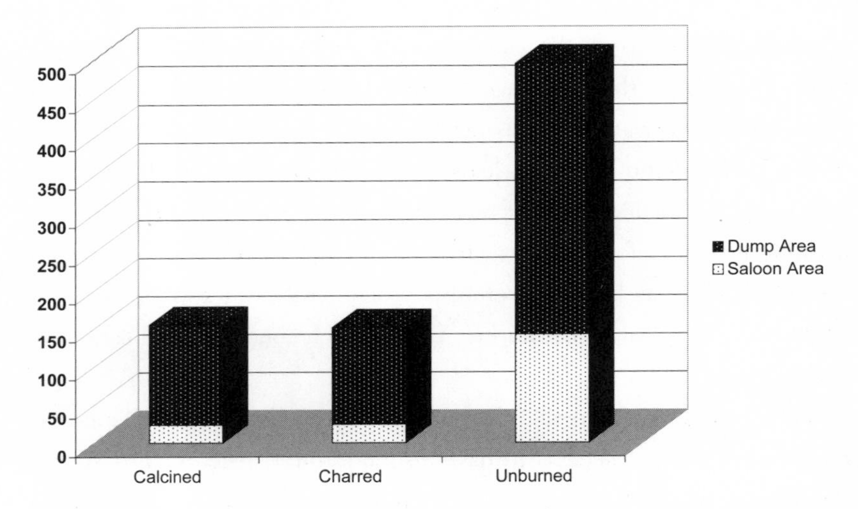

GRAPH 7.1. Frequency and distribution of calcined and charred mammal bones across the dump and saloon areas of the Boston Saloon site. The bulk of calcined and charred bones came from the dump.

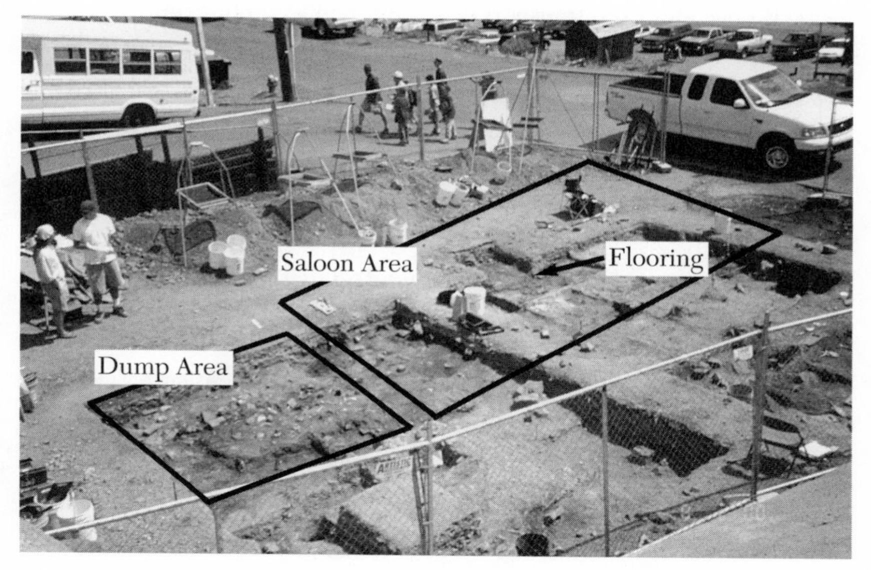

FIG. 7.1. Overview of the Boston Saloon site, demarcating the dump and saloon areas as identified by archaeological excavation.

that such an abundance of bones left unburned supports the argument that the 1875 fire was not responsible for scorching the entire dump area. Rather, the calcined bones were certainly the result of intentional incineration.

This scenario reveals a powerful detail about life on the D Street urban landscape in the vicinity of the Boston Saloon: It included the stench of burning animal remains and perhaps other garbage. While this may seem like a rather unpleasant image of a western boomtown, it underscores how we, with our penchant for sterile nostalgia, have lost sight of the more graphic details of urban life. Archaeologist James Deetz noted how such distorted views have been perpetuated by sanitary outdoor museum portrayals in colonial America. Calling on the power of archaeology, Deetz pointed out that the past was not necessarily "prettier, problems fewer, life simpler."[4] The image of burning animal bones and other refuse in the back alleys of western boomtowns may not paint a rosy, sentimental picture of the

West. It does, however, transport us back to a nineteenth-century moment on one of Virginia City's streets. Even though this moment included the sight of smoke rising from alleyways and the odor of burning garbage, it also contained the relatively pleasant aromas associated with lamb-based meals that pervaded the D Street landscape as they wafted out of places like the Boston Saloon.

Forensic Residue Analysis

The study of bone damage, such as calcining, and the distinction between human and animal bones represent rather traditional forensic treatments of archaeological materials. Other forensic techniques, such as chemical analyses using a gas chromatograph–mass spectrometer (GCMS), may not be as widely associated with archaeological studies, but they do have value in identifying biological residues clinging to artifacts from historic sites.[5] Forensic scientists use the GCMS to identify the chemical composition of substances recovered from crime scenes. While this is used on a daily basis in forensic labs to help solve crimes, it also has the potential to help solve mysteries at archaeological sites by identifying residues on objects left behind at historic places.

Archaeologists found residue on an object from the Boston Saloon and were able to put the GCMS to the test. Several pottery shards from at least two large matching stoneware crocks lay scattered around the dump area in the alleyway behind the saloon. Back in the lab, where the shards were sorted, cleaned, and prepared for mending, volunteer lab crew members noticed an unusual brownish-red stain on one of the lid fragments (figure 7.2).[6] One of the volunteers worked in a molecular biology lab and noted that the stain had the appearance of dried blood.

Because the substance initially looked like a bloodstain, it became a prime candidate for DNA testing, another useful forensic tool that will be discussed in more detail later. Before commencing with the rather costly DNA tests, however, the forensic lab needed to determine if the substance was actually

FIG. 7.2. Stoneware crock shard with a brownish-red stain that was subjected to forensic residue analysis. Photo by Ronald M. James

blood. The resulting analysis revealed that it was not blood—but this did not stop the persistent crime scene investigators and the eager archaeologists. Rather, it inspired them to seek other techniques, in the form of GCMS testing, to identify the source of the mysterious stain. Brainstorming about the stain's possible origin centered on the probable use of the crock. Since crockery was often used in kitchen contexts, it was clear that the object could have been associated with food preparation or food storage. Perhaps the stain was a remnant of a food product. The archaeological record helped the forensic lab to focus this speculation, for a number of pepper sauce bottles, including the rare Tabasco bottle, had also come out of this saloon's buried deposits.

Given the potential for red pepper–based residues at the Boston Saloon site, the forensic lab commenced GCMS tests for a substance called capsaicin in the stain. Capsaicin is extracted from the cayenne pepper, *Capsicum frutescens,* and is noted as the primary ingredient in McIlhenny's Tabasco Sauce; Worcestershire sauce similarly includes "chile peppers" among its ingredients, and Worcestershire bottles were recovered from the Boston Saloon as well. In addition to its popularity as a spicy condiment, capsaicin is a major ingredient of modern cayenne pepper–based chemical protection sprays used for self-defense. As a result of the presence of capsaicin-based spray residues in many criminal investigations, such materials are frequently examined in forensic cases through the technique of GCMS, and the forensic literature outlines a method for extracting and identifying this ingredient.[7] The potential for extracting pepper sauce from the mystery stain subsequently benefited from this existing research on capsaicin.

During GCMS examinations, the stain tested positive for capsaicin, suggesting that it did indeed represent a red pepper–based substance that probably came from a pepper sauce. The GCMS tests did not recognize capsaicin in other pepper-based products found at the Boston Saloon, such as Worcestershire sauce, indicating that the stain likely represented a product more like Tabasco sauce. The GCMS also picked up traces of acids associated with animal fat, results that gave the stain a "smoking gun" status regarding the use of a red pepper sauce on at least one meat-based menu item at the Boston Saloon. The stain therefore provides a rare archaeological remnant of leftovers associated with this saloon's cuisine.

DNA Analysis

Other biological evidence lingering on artifacts may not be as visible as the red pepper sauce stain. As a matter of fact, some of the most intriguing evidence is invisible to the naked eye. The "molecular blueprint" for all life on earth, deoxyribonucleic acid (DNA) can actually survive in fragments on archaeological sites, and bioarchaeologists have been attempting to extract ancient DNA (aDNA) for nearly two decades.[8] DNA becomes aDNA after a dead organism has begun to decompose or when biological residues—such as blood, saliva, sweat, and semen—are left behind to decompose on certain materials. Under ideal preservation conditions, such as freezing or desiccation, aDNA was initially given a life span that reached back millions of years; however, more recent and more conservative aDNA life spans appear to cover temporal periods of less than 100 years to perhaps as much as 100,000 years.[9]

Whether recent or ancient, DNA molecules carry hereditary information about gender and ancestry. If those molecules are preserved on certain artifacts from historic sites, which are well within the life span of aDNA, DNA tests on those artifacts can help archaeologists determine who came into contact with those items. With this in mind, the search began for Boston Saloon artifacts that appeared to be good candidates for maintaining DNA profiles at least 125 years old.

Buccal, or cheek, cells contain high concentrations of DNA; consequently, artifacts that would have come into contact with those cells became prime candidates in the search. A clay tobacco pipe stem fragment marred with teeth clench marks appeared to be a good sample, since the marks blatantly indicated that this object had made contact with the inside of someone's mouth (figure 7.3). The porous clay of the pipe stem also provided tiny catchments that might harbor DNA molecules over time. And finally, the pipe stem's bore hole inadvertently protected those molecules from ultraviolet (UV) light, which affects the molecular structure of DNA molecules.[10]

Forensic scientists carefully swabbed around the teeth clench marks near the bore hole, trying to lift DNA from that area of the pipe stem. Then they placed the microscopic sampled material in a solution with the chemical

FIG. 7.3. Stained white clay tobacco pipe stem with teeth clench marks at one end (right side of the object in this image); the marks made this a good candidate for DNA analysis. Photo by Ronald M. James

CHLEX, which binds to DNA. One nuclear DNA profile appeared. Upon looking closely at the profile, researchers saw that it contained two X chromosomes, indicating that a woman had held this pipe in her mouth more than 125 years ago.[11] This evidence provides unequivocal proof that a female was associated with at least one tobacco pipe from the Boston Saloon.

Forensic laboratory technicians then conducted calculations for people of various genetic backgrounds using allele frequency and distribution data as a means of determining the ancestry of this woman. Despite the potential for making determinations about her ancestral background from the DNA profile,[12] the tobacco pipe user's profile could not be associated with any one population group. While this result precluded identifying ancestral ties using the DNA profile, historical sources on African Americans in leisure contexts in the West suggest that women in such contexts were more likely to be of African than European ancestry.[13]

Linking Forensic Science with Vices and Saloon Archaeology

The synthesis of historical, archaeological, and genetic records helps us make microscopic but tangible contact with the person who held a tobacco pipe in her mouth while participating in the activities of an African American saloon.[14] This provides an incentive for rethinking the stereotype of the western saloon as white and dominated by males.[15] It also circles back to the topic of vices, drawing attention away from women's participation as servers of men's vices and to the idea of women as active participants in indulgences of their own.[16] Even a most respectable woman will engage in such activities if she likes them—and many respectable women did.

The existence of vices such as the use of morphine and laudanum by nineteenth-century women is relatively well documented. For example, the majority of morphine injectors were women from all walks of life.[17] A wide range of women also took laudanum, a product prescribed for a variety of ailments.[18] Laudanum was composed of a mixture of alcohol and opium and was introduced to European therapeutics in the sixteenth century; by the nineteenth century it became a common item included in the medicine kits of many "proper" Victorian families. Many women subsequently developed private, discreet opium habits.[19]

Such vices, especially the use of laudanum by women, were romanticized by Hollywood in a manner similar to, but on a lesser scale than, the sensationalism of western saloons.[20] Images of women smoking tobacco pipes are just not as popular[21]—or at least they are not as well remembered or as glamorous as images of women associated with more-perilous vices such as morphine injection and laudanum use. Either way, the paucity of women shown with tobacco pipes in our collective memory of western saloons, and especially in our collective memory as fabricated by Hollywood, underscores a gender bias with regard to this rather common indulgence. A fellow archaeologist made this clever remark about the female DNA on the Boston Saloon pipe stem: "You don't have the 'smoking gun,' you have the 'smoking pipe'" of women using tobacco pipes in western saloons.[22]

Techniques associated with forensic science, especially DNA recovery, can assist archaeologists in finding other "smoking guns" by helping them identify the kinds of people who handled and used the artifacts they unearth. The various beverage bottles, tobacco pipes, glassware, and ceramic pieces that dominate the Virginia City saloon artifact collections can certainly help us understand the ambience of those places. They can also help us understand the socioeconomic differences among the various establishments as revealed by the differences in elegant glassware, sophisticated interior fixtures, and high-quality meat cuts from the various drinking establishments. However, without oral histories to tell the stories of people who used those everyday saloon objects, most of the artifacts recovered do not easily lend themselves to ethnic or gender-based interpretations. As shown by the tobacco pipe fragment, remnant DNA can be lifted from certain objects, enhancing the potential for such specific types of interpretation possible.

This represents the beginning of a cooperative relationship between historical archaeology and forensic science. Forensic techniques have showed their vast potential for revealing unexpected, nearly invisible information from saloon artifacts. They provided an example of leftovers from a meal served in the Boston Saloon and added the unpredicted image of a woman smoking a tobacco pipe in that establishment. Artifacts have a unique way of creating a tangible connection with the past, and biological remains on those artifacts help us actually make contact with the people who last touched them, demonstrating how hard science can be integrated with the "humanistic science" of historical archaeology.[23]

CONCLUSION: CASTING THE SALOON OF THE WILD WEST IN A NEW LIGHT

Myth Versus Reality

The western story has been told most powerfully, most frequently, and most accessibly on film, providing mass audiences with an image of the Old West that they may assume represents the actual historical events of this region.[1] Western films and television shows have cultivated a widespread awareness of our western heritage and have generated saloon images in our collective memory that have their genesis in the false reality of movie sets.[2] Because the human right brain has the ability to intuitively create an entire reality from a few images,[3] mention of saloons in the American West tends to conjure up Hollywood-induced imagery about cowboys, gunfighters, and brawls. As a result, the western saloon has, for the most part, become associated with vice and violence.

The origins of such glamorized violence date to the early days of nineteenth-century mining boomtown development when people like Mark Twain openly admitted that, in his case, as a city editor for the *Territorial Enterprise,* he "let fancy get the upper hand of fact too often when there was a dearth of news"; Twain further confessed his glee on a slow news day when a desperado killed a man in a Virginia City saloon.[4] It is clear that on certain occasions violence was glamorized because it sold newspapers. Consequently, nineteenth-century media audiences were fed a wilder West than the one of reality.

So how can archaeology help us adjust the inaccurate picture of violence in western saloons? Along with an abundance of beverage bottles, artifacts such as tobacco pipes, dice, poker chips, a cribbage board, and dominoes answer this question, demonstrating that saloons were primarily places for people to amuse themselves while enjoying a drink and relaxing with a smoke. Vices or not, drinking, smoking, and gambling served as diversions from work and the hardships of life. Given the leisure connotations of such artifacts, their presence reminds us that people attended saloons for the most part to relax and socialize, not to find or cause trouble. Granted, gunfire was the primary amusement at places such as O'Brien and Costello's Saloon and Shooting Gallery, and it could certainly be interpreted as a violent form of recreation (figure 8.1). All the same, the shooting was usually aimed at targets in that establishment rather than at fellow drinking house patrons.[5]

Although less frequently than depictions in the violent Hollywood portrayals of life in the West would suggest,[6] guns were fired in and around some

FIG. 8.1. Cartridges from O'Brien and Costello's Saloon and Shooting Gallery. More than 1,000 casings were excavated at this site. Archaeologists found munitions at the other saloons, but in much smaller quantities. Photo by Ronald M. James

Virginia City saloons that were not shooting galleries, as indicated by archaeological evidence of ammunition cartridges from revolvers and rifles. The reasons these items wound up in the saloon deposits are unknown. Some reports satirize revolvers as a common accessory that men wore in many northern Nevada communities,[7] and the presence of munitions from those handguns shows that they were being fired once in a while.

At least one explanation exists for the accidental discharge of a revolver at the Boston Saloon. A group of men, each with a pistol resting in his lap, sat around a table playing a "friendly" game of poker just before midnight on an August evening in 1866. As the game went on, a derringer slid from someone's lap, hit the floor, accidentally fired, and shot one of the poker players in the calf of his left leg, taking out a piece of bone. The victim of this mishap was a man named Frenchy, who was the only white man in the saloon at the time. He apparently got along quite well after receiving medical attention.[8]

The accident provides an example of the countless dimensions of "wildness" in the West that perpetuated the belligerent stereotype of life in that region. Such mishaps were probably commonplace in a town where many people carried and wore small firearms, but they do not necessarily mean that violence abounded in saloons. Rather, stories like the one about the Boston Saloon highlight the myriad ways in which violent behavior became so overstated and partially explain the occasional presence of bullet casings in the archaeological remains of the saloons under study. As fighting and hostility were sensationalized, in historical reports and later in Hollywood portrayals, saloons and western boomtowns became sites of violence in our collective memory.[9] While this rendition of western history flourished, the real moments of living and being in a boomtown environment tended to be overlooked and forgotten. History and archaeology have the potential to bring those moments back and to help us understand the distinctions between popular caricatures and realism with regard to the western heritage. The remains of each saloon allow us to step back into those instants, to pass through their long-deteriorated doors, and to experience the individual personality of each place.

Public Drinking and the Time Line of Human History

For all the variety among saloons, the activities in such establishments—namely drinking, socializing, game playing, and relaxing—were quite similar. Ironically, they were not dramatically different from either those showcased in Hollywood films or those engaged in in drinking houses throughout history. People have been enjoying public drinking places for a longer span of time and over a wider geographical range than the nineteenth-century American West. If the western saloon is considered on a broader time line of all of human existence, it becomes possible to step back and take a look at a bigger, global picture of social drinking in public places and to understand that this practice did not originate in the American West during the nineteenth century.

As a matter of fact, people have been sharing drinks together for thousands of years, as shown by one of the earliest known forms of documentation: cylinder seals from Mesopotamia, the ancient region in the Near East that sprawled across the Tigris and Euphrates river valleys. These objects were made of materials such as shell, lapis, hematite, and serpentine and were engraved with scenes and symbols used to authenticate written clay records, letters, and proprietary rights.[10] The artistic carvings that personalized each seal represented a graphic response to the world that has become a valuable visual record for us today, accentuating the antiquity of many activities, such as social drinking.[11]

Many of those carvings illustrated ritual banquets, which appeared on cylinder seals, plaques, and friezes in Mesopotamia by about 2340 B.C.E. In many of these artful banquet scenes, participants used long straws or tubes to drink beer from large communal vessels that sat on the ground (figure 8.2). The straws or tubes allowed access to the ale that lay beneath the scum on the beverage's surface, and this was the usual manner of enjoying beer in

FIG. 8.2. The design in the upper left portion of this seal imprint shows a pair of seated figures sipping ale from a large vessel through pipes or straws; the seal itself dates from 3000 B.C.E. © Copyright The Trustees of The British Museum; see D. J. Wiseman, *Cylinder Seals of Western Asia,* plate 24 (London: Batchworth, 1958) and H. Frankfort, *Cylinder Seals: A Documentary Essay on the Art and Religion of the Ancient Near East*, 75–77 and Plate XIVf (London: Macmillan and Co., 1939).

the ancient Near East.[12] Metal strainers were fitted into the bases of the tubes and straws, and archaeologists have discovered these while working in Mesopotamia, Syria, and Egypt.[13]

In addition to sharing strainers with the tube-style drinking of Mesopotamia, Egypt had its own array of public drinking houses, known as "houses of beer."[14] The ancient city of Pelusium, which used to sit along the easternmost branch of the Nile, was well known for both its university and its proliferation of drinking houses.[15]

The Egyptians introduced beer brewing to the Greeks, who probably exported it to the Romans.[16] Refreshment taverns or *tabernæ,* sprang up along Roman roads throughout Europe. Taverns appeared in England by the first century A.D., providing lodging, food, and drinks, such as wine and ale, for travelers.[17] Eventually, various types of inns or *tabernæ* came to reflect Europe's cultural diversity: "the wine shop of the Mediterranean coast, the German beer hall, the café and cabaret of France, and the English public house, or pub."[18]

Across the Atlantic, the American version of the European inn descended from and was similar to the English pub. By the revolutionary era and the early Republic period these establishments were referred to as "taverns" and provided meals and lodging for travelers and isolated residents in rural areas. They became places dedicated solely to social drinking in American cities.[19] These taverns were "far more than places to imbibe," however; they also became places for men to read newspapers and discuss political or public issues.[20]

Regional cultural historian Elliott West found that American taverns took on various characteristics that reflected the socioeconomic levels of their patrons, with terms such as "grogshop," "groggery," and "doggery" used to describe the less-reputable establishments. Intending to impart a more respectable air to these institutions that increasingly received criticism from community moral leaders, business owners began to use the term *saloon*. This was derived from the French *salon* and was used by English speakers in both England and America during the early eighteenth century to describe a large room where public meetings and events were held. However, the term took on more aristocratic connotations in America, giving the feeling of "gentility and restraint" to public drinking places, as opposed to the view of their patrons as drunken disturbers of the peace.[21]

The word *saloon* was in common use by the mid-nineteenth century, appearing frequently in major city directories and becoming a generic designation for a variety of places where alcohol was sold and consumed.[22] Despite its initial images of refinement and proper behavior, by the end of the nineteenth century, the term had come to "conjure up visions of vice, filth, and slobbering drunkenness."[23] By this time saloons had already appeared all over the western mining landscape. This environment subsequently became the standard setting for such drinking establishments in our mainstream memory. As a result, the saloon is a powerful icon of America's western history.[24]

Diversity and Virginia City Saloons

This study compels us to take a new look at this national symbol and offers insight into the cosmopolitan makeup of western boomtowns. By catering to distinct groups, saloons remind us of the ways in which this nation's diverse heritage developed over time. Because of its melting pot status as a nation "literally made up of every part of the world," the United States has valorized multicultural diversity as a significant modern issue.[25] One scholar calls attention to the irony of this position, since heterogeneous nations such as the United States originated in "conquest, slavery, and exploitation of foreign labor."[26] It is important to consider the historical developments of diversity, then, as many such nations were based on antagonism toward others. The historical and archaeological records of western saloons reveal at least some insight into such developments.

Because of the socioeconomic and ethnic segregation prevalent in the population, diversity among saloons was common and signaled the final stage of a mining camp's transformation into a city.[27] The population of mining communities increased by virtue of an influx of people from all over the world, which amplified the cultural and ethnic diversity of boomtowns. This array of people from various ethnic and cultural groups came into contact with each other in cosmopolitan western boomtown settings,[28] and the new immigrants to the West sought to "soften the blow" of the anxiety and hostility associated with the transition to a new life, partly by actively expressing their identity in various leisure venues.[29] Prejudicial treatment, which likely inspired people's tendency to spend their leisure time with others of similar background, ultimately influenced saloons' diversity. As leisure institutions, saloons were physical places where people could socialize and relax with others like themselves.[30] Like other leisure establishments, saloons subsequently encouraged segregation and pluralism instead of "indiscriminate social mixing,"[31] and ultimately the assortment of saloons in a mining boomtown reflected the town's diverse social, cultural, economic, and ethnic milieu.

What, then, do the archaeological remains and the historical records of Virginia City saloons tell us about the development of diversity in the mining West? While different levels of sophistication among the various saloon artifacts illustrate how the archaeological record reveals socioeconomic diversity among those establishments, artifacts do not clearly indicate ethnic or cultural diversity among saloons. Rather, historical records, including census manuscripts and directories, reveal that kind of detail, providing such information as the names of the various saloon proprietors and their ancestral backgrounds. By linking the saloon owners with the socioeconomic distinctions revealed by the items recovered during archaeological excavation, it becomes possible to make some statements about diversity in this boomtown.

For example, Piper's Old Corner Bar boasted the most opulent interior decor and fixtures of the four saloons, illustrating its place at the classy end of Virginia City drinking establishments. Objects such as the decorative meerschaum tobacco pipe indicate that this bar's owner or patrons had finely crafted personal possessions and probably were financially comfortable. Those patrons were described as "gentlemen."[32] This historical description, along with the scarcity of artifacts associated with women, drives home the point that Piper's maintained a rather exclusive, masculine identity. The artifacts, however, do not give any indication of the ethnic backgrounds of the gentlemen who frequented the establishment, although it is assumed that they were European Americans and, like John Piper himself, European immigrants. While there are no known incidents of northern Paiutes, Mexicans, overseas Chinese, or African Americans trying to enter this upscale saloon, those individuals were treated with prejudice elsewhere in Virginia City's daily life,[33] which means they were probably not going to seek refuge and camaraderie in a place like the Old Corner Bar.

Historical records do, however, suggest that people of African ancestry patronized the Boston Saloon.[34] As a matter of fact, the mere existence of this business underscores an African American presence in the region, reminding us of a major aspect of the African American heritage: the most significant unforced migration of people of African descent occurred during the latter portion of the nineteenth century, after the California gold rush.[35] This represents one of several facets of the mining West that for the most part have been excluded from many portrayals of western history.[36] Inspired by word that there was more tolerance in the West for people of color, African Americans moved westward hoping to elevate their social status and to reach the "same economic starting point as others in America."[37]

As the primary social drinking house serving people of African ancestry, the Boston Saloon probably did not cater to one particular socioeconomic group. Rather, because of the relatively small population of African Americans living in Virginia City and the paucity of saloons that catered to them, this was most likely a place where well-off individuals mingled with people who worked in low-paying jobs. This is only an assumption, though, made on the basis of the variety of occupations held by African Americans in Virginia City and the prejudicial treatment they probably received at other saloons, as implied by one person's lament for "a place of recreation of our own" and by stories of African American writers elsewhere in the West.[38]

Archaeological finds from the Boston Saloon indicate that its patrons experienced an upscale atmosphere. This establishment was therefore distinct from the other Virginia City saloons because of its ownership and clientele and because it was among the classier saloons in that community. Such a fancy saloon likely was a source of pride for African Americans, given the concurrent contexts of prejudice and optimism associated with the post–Civil War Reconstruction era.[39] This kind of ambience also challenged historical stereotypes of African American saloons as seedy criminal establishments.[40] Likewise, it contradicted modern stereotypes, which were observable in the shocked reactions of numerous visitors watching archaeologists brush dust off sparkling stemware from Virginia City's African American saloon.

It is possible that such high-class goods were used by William Brown in his saloon in order to combat racist assumptions about an African American way of life.[41] Indeed, the frequent instances of prejudice in Virginia City's boomtown setting warranted the effort to resist such assumptions.[42] On the other hand, Brown had actually attained success as a result of his entrepreneurial endeavors, and perhaps the saloon fixtures merely reflected his relatively high socioeconomic standing rather than his active attempts to fend off racism.[43] Regardless of their backgrounds, many people who profited from the mining wealth of Virginia City displayed their status through food delicacies and material goods, including ornate architecture that still stands. In this sense, William Brown was just like everybody else.

The archaeological remains from places like O'Brien and Costello's Saloon and Shooting Gallery told another story. Two Irishmen, O'Brien and Costello, collaborated to open the combined saloon and shooting gallery in the underworld section of Virginia City at the climax of the community's mining boom. Despite the Barbary Coast's seedy reputation, O'Brien and Costello attempted to outfit their business with a handful of fancy fixtures, such as glass decanters and a few pieces of crystal stemware. These objects add another dimension to our understanding of the Barbary Coast by suggesting the ways in which at least one of the so-called dodgy saloons in that neighborhood sought to spruce up its interior ambience. Whether such materials mitigated the questionable character of the saloon is certainly debatable, and why the Irish saloonkeepers chose to use them is even more of a mystery. Perhaps they were trying to use material culture as a means of combating racist, classist perceptions of Irish people.[44] In doing so, they showed others that even an Irish-owned establishment from a low-rent area could boast sophisticated furnishings.

Building on this story, clothing accoutrements from fancy women's apparel indicate that rather well-dressed women were associated with this shooting and drinking business. Given the Barbary Coast's scandalous prostitution stories, the knee-jerk interpretation of those items could be explained as evidence of a prostitute. While this is entirely possible, it cannot be proved. Nonetheless, the artifacts at least tell us that women wearing fine dresses and dainty boots were somehow involved in the activities of O'Brien and Costello's establishment.

Even amid the underworld, it was possible to experience at least a few elements of material refinement. Nevertheless, a few pieces of stemware and some women clad in fancy clothing did not likely make the Saloon and Shooting Gallery more respectable in the eyes of the reputable members of the community; it was after all, located in a shady part of town. This means that O'Brien and Costello, despite their attempts at improving the experience of their business, still did not attract Virginia City's high-society patrons. In other words, if a wealthy, non-Irish patron, such as an Englishman or a Cornishman, had entered, he might have been served but he probably would not have felt welcome.

Nearly ten years after O'Brien and Costello's operation served a slice of Virginia City society, the Hibernia Brewery opened up the outskirts of a "cleaned-up" Barbary Coast.[45] The name chosen by Shanahan and O'Connor, the two men who started the establishment, clearly referred to an Irish heritage. This Irishness became even more vivid when the bale seal with a lyre design from Ireland emerged from the excavation trenches. Shanahan and O'Connor were Irish men living in America, as were many of their customers, and this minute item provided all those who saw it with a reminder of their heritage, and perhaps their identity, in a boomtown far from their homeland. The display of such an emblem suggests that their dignity prevailed despite the climate of hostility and prejudice that Irish-descended people experienced in America during the first portion of the nineteenth century.[46]

Yet the materials left behind from the Hibernia indicate that it was a sparsely decorated establishment. When viewed in the context of all four of the Virginia City drinking houses studied here, the Hibernia's unexceptional artifact collection yielded some information. With its basic, utilitarian fixtures, this saloon fit within the group as a rudimentary place dedicated to public drinking with few to no frills. The paucity of women's artifacts at the site adds more to this story. Perhaps the Hibernia was another sort of "gentlemen's club," one that may have catered to slightly poor but God-fearing and respectable Irish men with families, who just wanted to relax, socialize, and have a drink with other men who shared similar lives and worldviews. Certainly the Hibernia's doors were open to non-Irish people, and the place likely catered to many other Europeans and European Americans at the lower end of the socioeconomic scale. However, given the potential conflicts with English people[47] and the racist treatment of African Americans by the Irish as they attempted to embrace their "whiteness,"[48] those groups likely avoided this saloon because they probably would not have felt welcome there. Not at all unlike African Americans in the Boston Saloon, the Irish created solidarity among themselves and found a haven from prejudicial treatment in places like the Hibernia Brewery.

During the first half of the nineteenth century, more than three million Irish immigrated to America to escape the Anglo-Irish Anglicans. They basically traded one anti–Irish Catholic group for another—WASPs—once they reached American soil. In their new country they found themselves being discriminated against and being treated as "black"; however, in the years before and after the Civil War, the Irish "became white" by embracing racism against African Americans.[49] Some of Virginia City's most successful and famous mining barons, such as John Mackay, were Irish,[50] proving that it was possible for the Irish to overcome poverty, and perhaps prejudice.[51] Mackay and the others were fortunate, shrewd, and happened to live during a time when the Irish were making a transition to a higher standing in American society. Archaeological remains from places like Shanahan and O'Connor's Hibernia Brewery are perhaps reminiscent of the poverty-stricken Irish who were still experiencing the stigma against their people during this transition to whiteness. On the other hand, the nicer fixtures from O'Brien and Costello's Saloon and Shooting Gallery may represent attempts at using material culture to rise above the prevailing stereotypes at that time.[52]

This is not to say that there was an absence of stigma against African Americans. On the contrary, their lives were by no means as easy as those of their white neighbors. Racist undertones and overtly restrictive attitudes and laws affected Virginia City's African Americans, making their lives a complex juxtaposition of integration and prejudice, of neighborly acceptance and ill treatment.[53] Such variation in treatment was also experienced by African Americans living elsewhere in the West.[54] The Boston Saloon illustrates a success story in this volatile context.

With its success came a unique atmosphere, including music that was not necessarily from a clichéd saloon instrument: the piano. While evidence of piano keys was recovered from Piper's Old Corner Bar, archaeologists found a trombone mouthpiece at the Boston Saloon, serving as a reminder of the different kinds of entertainment offered by various saloons and enhancing the distinct ambience of the Boston Saloon amid Virginia City's public drinking houses (figure 8.3).[55]

The story of William Brown and his Boston Saloon highlights the existence of a nearly forgotten remnant of African American heritage in the West, and its place on the comparative scale of Virginia City drinking houses leads to an understanding of the dimensions of diversity. The Boston Saloon and Piper's Old Corner Bar just happened to be places with atmospheres exuded an association with the upper echelons of Virginia City's socioeconomic hierarchy, while O'Brien and Costello's Saloon and Shooting Gallery and the Hibernia Brewery were more probably the choice of the lower end. This insight, garnered from the comparison of artifacts, reminds us that diversity should not always be boxed into categories that dance

FIG. 8.3. Brass mouthpiece from a trombone from the Boston Saloon.

around differences defined by major cultural groups. Rather, it is clear that numerous socioeconomic factions were also involved, making for even more complex levels of variation within each of those groups.

Some artifacts even indicated women's participation in the multifaceted mix of saloons, which deepens our understanding of yet another, gender-based component of the archaeological record and of the past in general.[56] In the case of Virginia City, attention to gender helps to expand the dimensions of saloon diversity. For instance, women were definitely participating in the activities of the Boston Saloon and O'Brien and Costello's Saloon and Shooting Gallery, but they were not as much in evidence at Piper's Old Corner Bar and the Hibernia Brewery.[57]

Then again, the distinction may reflect the taming influence of middle-class values on the "wild" West. Middle-class ideas of morality and "ladyhood," which European American women, usually miners' wives and families, brought with them to western mining camps by the 1850s, prohibited respectable women from associating themselves with saloons.[58] Drinking houses were public places where men gathered, with drinking in the late nineteenth and early twentieth centuries becoming one of the most gender-segregated activities in the United States.[59] For the most part, the women who were involved with such male-oriented, vice-laden settings were perceived as having questionable reputations themselves—whether they actually warranted such reputations or not.[60] For this reason, well-thought-of saloons, such as Piper's Old Corner Bar, were probably frequented by gentlemen, as implied by historical documents and as emphasized by the absence of artifacts associated with women. Undeniably, many of those gentlemen sought female companionship, but they most likely had to patronize other businesses to get it. This meant that places like Piper's remained relatively highly regarded, while others, such as O'Brien and Costello's, were rendered less reputable.

On the other hand, the apparent scarcity of women at the austere Hibernia Brewery reminds us that the story cannot be that simple. Much like Piper's Old Corner Bar, few to no women were archaeologically visible in the artifacts from that Irish-owned business. This implies that the Hibernia Brewery catered primarily to men—perhaps men of Irish ancestry. Despite the active role of women in other Irish social drinking contexts,[61] they seem to have been scarcely involved at all with the everyday operations of this establishment. Although the reasons for this may never be explained, it is possible that the Hibernia was simply a pub, without the posh interior decor of Virginia City's well-known gentlemen's saloons, where Irish men could go out for a drink without being in the presence of women with questionable reputations.

Women at the Boston Saloon add yet another twist to our understanding of gender and diversity in saloons. Unlike Virginia City's European Americans and European immigrants, people of African ancestry did not have the wide choice of respectable, nonrespectable, classy, or seedy drinking establishments. Consequently, women in the Boston Saloon, like the men, probably represented various socioeconomic classes. Certainly many African

American women, in the name of self-improvement, respectability, and temperance, avoided the place.[62] Other women, such as J. Lind, may have worked there as courtesans. Still others, such as Amanda Payne, most likely frequented the D Street enterprise for business purposes.[63]

The DNA on the tobacco pipe stem takes this story a step further by linking a woman with a common indulgence at the Boston Saloon. The image of a woman smoking a pipe in an African American saloon challenges prevailing assumptions of who was relaxing in public drinking places in the West. Without a doubt, this is a unique situation, with information made possible by the serendipitous convergence of historical research, archaeological excavation, DNA preservation, and advances in forensic science.[64] Nevertheless, it should inspire everyone who reads this book to abandon their stereotypical assumptions about western saloons and open their minds to the innumerable accounts that make up our complex history. People from various cultures, classes, and genders participated in the enduring activity of social drinking, creating vast pockets of wide-ranging saloon experiences throughout urban mining boomtowns. A view of each saloon as a slice of a shared, diverse history in the mining West makes it clear that these are merely segments of a multifaceted narrative that help us understand the bigger historical picture of the "winning" of the West.

History and Archaeology Go Hand in Hand

This saloon study illustrates the ways in which historical archaeology can enlighten our view of the past. A larger issue still looms, however, in regard to the discipline's obligation to address questions such as why dig up literate societies when the same information may be available in written records?[65] In any consideration of this question, it is necessary to think about who was literate and therefore who was doing the writing in the past. Not everyone was literate, which means that relatively few people documented history. Any individual writer will portray an event with his or her own bias or worldview; thus, depictions of even well-documented historical events communicate a one-sided perspective. In the case of saloons, past writers exaggerated some events, and adding the problems associated with the survival of certain records increases the possibility for biased depictions.[66] This is not to say that historical records cannot be trusted. Quite the opposite, in fact. Historical records provide some of the richest insights into the past. Such records provide the general infrastructure of past events, but it is necessary to consider all the people who are missing or who are underrepresented from such reconstructions of our history.

Accordingly, historical archaeology's most significant contribution has emerged from its ability to make pasts for those people who were not so well documented by the limited perspectives of the historical records.[67] The recovery of archaeological remains left behind by those people may, in some cases, be our only way of knowing anything at all about their lives in certain places and at certain times. Ideally, this information can be used to enrich history and can help to give us a better understanding of recent development on our human time line.

Such an understanding could cause us to think more deeply and reflect for a moment the next time we are socializing and enjoying a vice, about how people have been participating in these activities for a long, long time, whether leisure as we know it took place in certain locations such as saloons or was simply woven into people's daily lives.[68] This may inspire us to consider how and why we socialize with certain people at particular places. This, in turn, adds depth to an understanding of ourselves as well as saloons.

Popular portrayals of public drinking places have overlooked their information potential in the name of romanticized, sensational imagery. Historical archaeology has helped to deconstruct this popular segment of American history while desensationalizing the collective memory of the "Wild West."[69] This investigation has demonstrated how saloons and their seemingly humble assortment of lost, broken artifacts provide minute, particular details that help create new ways of viewing boomtowns, saloons, and life in the West, with its cosmopolitan mix of people.[70]

This aspect of western history is not really very far removed from the present, which means that this historical and archaeological study of saloons can provide a relevant link to contemporary modern issues.[71] The existence of an African American–owned saloon, the different types of Irish-owned saloons, the German-owned opera house saloon, and the participation of women in certain businesses remind us that diversity was as prevalent then as it is today. Furthermore, this study illustrates just a few of the dimensions of our collective heritage in the mining West, which should open the floodgates for developing an even broader understanding of the diversity that has been present in the region through Native American groups, Spanish explorers, Mexican miners, Chinese railroad workers, African American entrepreneurs, Irish laborers, middle-class Anglo-American housewives, and so on. An awareness and understanding of this common history may, in turn, help to dissipate prejudice in modern America by providing all Americans with a tangible link to a united heritage in this region. It is necessary to continue to promote and expand upon this story of a shared heritage in the West to highlight a sense of mutual respect for the diverse cultures composing the history and current character of this country.[72] History and archaeology go hand in hand to illuminate the details of such an intricate chronicle between the past and present.

Fade to the Present

Although the value of such "hand-in-hand" research may seem obvious here, that is not always the case. Usually people's introduction to historical archaeology sparks a number of common reactions: (1) we already know enough about the recent past; (2) we can get that information from history books; or (3) this is an expensive way to study history.[73] But there are still massive discoveries to be made by inquisitive archaeologists who combine their penchant for digging with an examination of the historical documents. Even though our history within the past few centuries encompassed the lives of people who were relatively similar to ourselves, we are still worlds apart.

Together, history and archaeology help us make contact with those individuals by revealing some of their names, by allowing us to touch the objects that they handled, and by letting us, in this case, step into their saloons.

While many boomtown saloons have long settled into dust and ruin, some are still standing and still serve drinks. One does not need to draw upon history and archaeology to experience these places. Rather, it is possible to understand the perspective of a modern saloon patron by simply driving up a winding mountain road to Virginia City, Nevada, sauntering down the creaking boardwalk, and passing through the doorways of several saloons that are interspersed among rows of historic building facades. The old structures are constructed right next to one another, with brightly painted signs inviting tourists to sample fudge, try on cowboy hats, and slip into a cool, dark drinking house. For all the visual stimuli on this revived "ghost" town's streets, the large sign reading "Bucket of Blood Saloon," with its illustration of a spilling bucket of blood, beckons visitors to test their luck with the Wild West experience.

Most visitors inevitably stop and have a drink at the Bucket. Live banjo and piano music wafts out of this landmark's graceful double-door entrance, and on parched summer days misters spray refreshing clouds of water on all who pass the building's entrance. The bar sits just inside the entrance and extends for most of the length of the right-hand side of the building. To the left, rows of slot machines blink and make bleeping sounds. Straight ahead, a large window set into the building's back wall reveals the surrounding panorama of desert mountains; the Boston Saloon was once located behind the structure that houses the Bucket of Blood, which means people could look out that window and observe the archaeological excavation of a historic saloon while enjoying a drink in an operating saloon. Every space, from floor to ceiling, in the Bucket is filled with antiques. Exquisite lamps dangle from above, while oil paintings and historic photos are hung on every bit of wall. The gentlemen bartenders wear uniforms with starched white shirts, black-and-gold vests, bow ties, and ruffled armbands. On busy days there is standing room only around the bar, and those in the know order drinks like mint juleps.

By early evening, the sun slips behind Mount Davidson and most tourists leave in order to make their way down the precarious switchbacks toward their resort hotel casinos in Reno. As everything quiets down, murmurs and laughter rise from the locals relaxing in the various Virginia City saloons, which are pretty much the only businesses that stay open throughout the evening. Cigarette smoke and spilled beer emerge from each doorway along the boardwalk. As the last glow of daylight wanes, lights inside each place shine pockets of illumination along the street, interspersed with the shadows cast by modern versions of historic gas lamps.

This is a quieter time than the boomtown's evenings during the mining heyday, but it is still possible to sense the past by examining the area's history and archaeology and simply by just taking in Virginia City and the surrounding Comstock Mining District. History is alive here, and places like the Bucket of Blood and its neighbor, the Delta Saloon, gracefully hide their age under fresh paint, sparkling antiques, and the bells and whistles of slot machines. Many other saloons, wedged in amid the assortment of candy stores and gift shops, draw visitors with the charming odors of old buildings. The fragrance and feeling of bygone days live on in places like the Silver Queen, the Silver Dollar, the Union Brewery, and the Washoe Club. Still others, including the Bonanza, the Mark Twain, and Julia Bulette's, are inspired by well-known characters from television and local folklore.[74]

Playing on historical characters and using buildings and fixtures associated with the historic saloons that occupied their spaces, these "modern" drinking houses have their own identities. Aside from adventurous tourists, each business tends to attract certain segments of the small community still living in Virginia City. In their own ways, today's saloons offer a glimpse into the past, but it is an elusive vision. After all, they belong to their own century, and deciphering the modern identities of these haunts in an aged mining landscape is better left for another saloon story.

NOTES

INTRODUCTION ~ HISTORICAL ARCHAEOLOGY METHODS

1. The term *site* is a generic term for a grouping of artifacts or material remains and features that, collectively, demonstrate a concentrated area of human activity in the past. For the most part, it is common archaeological jargon and needs little more definition or discussion. However, some archaeologists argue that the notion of "site" should be discarded altogether in favor of smaller units of analysis: artifacts and the continuous distribution of those artifacts on and near the surface of the earth; for a discussion of this view, see Robert Dunnell, "The Notion Site," in *Space, Time, and Archaeological Landscapes*, edited by Jacqueline Rossignol and LuAnn Wandsnider, 21–41 (New York: Plenum, 1992); see also James Ebert, *Distributional Archaeology* (Albuquerque: University of New Mexico Press, 1992).

2. Special thanks to Giles C. Thelen for reminding me of the positive effects of archaeological excavation.

3. Larry McKee, "Public Archaeology," in *Encyclopedia of Historical Archaeology*, edited by Charles Orser, Jr., 456–458 (London and New York: Routledge, 2002).

4. James Deetz, Foreword to *A Chesapeake Family and Their Slaves: A Study in Historical Archaeology*, by Anne Elizabeth Yentsch, xviii–xx (Cambridge: Cambridge University Press, 1994).

5. Even with the basic methods, each archaeological project tends to treat its resources on a case-by-case basis, slightly modifying standard techniques to work with the nuances of the infinite number of site types.

6. For example, Patricia Nelson Limerick, *The Legacy of Conquest: The Unbroken Past of the American West* (New York: Norton, 1987); Elliott West, *The Way to the West: Essays on the Central Plains* (Albuquerque: University of New Mexico Press, 1995); Mary Martin Murphy, *Mining Cultures: Men, Women, and Leisure in Butte, 1914–1941* (Urbana: University of Illinois Press, 1997); Ronald M. James, *The*

Roar and the Silence (Reno and Las Vegas: University of Nevada Press, 1998); Ronald M. James and C. Elizabeth Raymond, eds., *Comstock Women: The Making of a Mining Community* (Reno and Las Vegas: University of Nevada Press, 1998).

7. Donald L. Hardesty et al., *Public Archaeology on the Comstock,* University of Nevada, Reno report prepared for the Nevada State Historic Preservation Office (Carson City: Nevada State Historic Preservation Office, 1996); Kelly J. Dixon et al., "The Archaeology of Piper's Old Corner Bar, Virginia City, Nevada," Comstock Archaeology Center Preliminary Report of Investigations (Carson City: Nevada State Historic Preservation Office, 1999); and Kelly J. Dixon, *"A Place of Recreation of Our Own." The Archaeology of the Boston Saloon: Diversity and Leisure in an African American–Owned Saloon, Virginia City, Nevada* (Ann Arbor, MI: University Microfilms International, 2002).

8. In addition to fieldwork, field school students spent time in the University of Nevada's Getchell Library and Special Collections, the Nevada State Library and Archives, and the Nevada Historical Society to experience historical research in relation to the site upon which they worked. They prepared notes of their historical research, many of which included maps denoting a reconstruction of businesses and other activities around the Boston Saloon site, to accompany their field notes. Finally, each student turned in a formal paper that addressed an aspect of archaeological fieldwork. Visiting scholars also gave Virginia City field school students, volunteers, and public visitors the opportunity to learn about the different ways in which various archaeologists approach field methods. For example, Adrian and Mary Praetzellis of Sonoma State University provided insights on the methods of conducting urban archaeology and interpreting the complex stratigraphic contexts of an urban site. During his visit, Paul Mullins of Purdue University gave a slide presentation detailing his work with African American archaeology in Annapolis; this gave students working on the Boston Saloon site an understanding of how their project fit in the bigger picture of African American archaeology.

9. The Boston Saloon Project's education director, Dan Kastens, orchestrated the visits of such large numbers of children, going to many of their schools before the field trips to the site to provide young students with brief overviews of the project and excavation methods.

10. The media were also a consistent presence during the five-week period of excavation. Media coverage of the Boston Saloon project spanned local print and television news, as well as a nationwide range of media outlets such as the *San Francisco Chronicle,* September 19, 2000; the *Boston Globe,* September 12, 2000; *Archaeology,* November/December 2000; and *American Archaeology,* Winter 2000–2001.

11. The committee included Michael S. Coray, special assistant to the president for diversity at the University of Nevada, Reno; Lucy Bouldin, director of the Storey County Library, Virginia City; Ken Dalton of the Reno-Sparks NAACP; Elmer Rusco, professor of political science, University of Nevada, Reno; and Theresa Singleton, Syracuse University, Syracuse, New York; finally, Lonnie Feemster, then president of the Reno-Sparks NAACP, received a copy of the research design. In addition, the Comstock Archaeology Center Technical Advisory Board (Ken Fliess, Don Hardesty, Gene Hattori, Ron James, David Landon, Pat Martin, Susan Martin, and Ron Reno) received copies of the research design for review before the fieldwork commenced.

12. Ron James carried out the initial historical investigations that were specifically focused on the Boston Saloon. These are summarized in "African Americans on the Comstock: A New Look," a paper presented at the Conference on Nevada History, in Reno, Nevada, May 1997. In *"Good Times Coming?"* Elmer Rusco provides a history of African Americans in Nevada during the nineteenth century and presents a starting point for understanding African Americans on the Comstock; *"Good Times Coming?" Black Nevadans in the Nineteenth Century* (Westport, CT: Greenwood, 1975).

13. The "West" here is defined as the area west of the 100th meridian, and the "mining West" is treated as the extant and ruined remnants of mine operations and boomtowns throughout that region. The literature on African American archaeology reveals many works on African American history in the West; for many African Americans living in the nineteenth century, the West was seen as a place of "economic opportunity and refuge from racial restrictions," yet African American writers' accounts of prejudice reflect the disillusionment that many African Americans experienced after confronting racist restrictions and attitudes. The first quote is taken from Quintard Taylor, *In Search of the Racial Frontier: African Americans in the American West, 1528–1990* (New York: Norton, 1998), 81, and the accounts of prejudice came from African American writers, such as Thomas Detter, and were published in the *Pacific Appeal*, February 22, 1868, 2, and October 8, 1870, 1. The *Pacific Appeal* was an African American–edited newspaper that operated out of San Francisco between 1863 and 1883. Its circulation reached various communities throughout northern California and extended to other areas, including Idaho, Nevada, Oregon, Washington, Victoria Island, and Panama. This resource contains an array of primary sources written by people of African ancestry, offering alternative perceptions of life in the West, rather than the more prevalent Eurocentric observations about that region. Despite these and other writings about African Americans from the standpoint of western history, there were only a few archaeological investigations of free black populations in that region: Todd Guenther, "At Home on the Range: Black

Settlement in Rural Wyoming, 1850–1950" (master's thesis, University of Wyoming, 1988); Adrian Praetzellis and Mary Praetzellis, "We Were There, Too": Archaeology of an African-American Family in Sacramento, California, Cultural Resources Facility, Anthropological Studies Center (Rohnert Park, CA: Sonoma State University, 1992); Margaret C. Wood, Richard F. Carrillo, Terri McBride, Donna L. Bryant, and William J. Convery III, *Historical Archaeological Testing and Data Recovery for the Broadway Viaduct Replacement Project, Downtown Denver, Colorado: Mitigation of Site 5DV5997*, Archaeological Report No. 99–308 (Westminster, Colorado: SWCA, Inc., Environmental Consultants, 1999); Adrian Praetzellis and Mary Praetzellis, "Mangling Symbols of Gentility in the Wild West," *American Anthropologist* 103 (2001) 3: 645–654. Thus, the archaeological record of African Americans was scant west of the 100th meridian and absolutely lacking in the context of the mining West.

14. James, *The Roar and the Silence,* 152.

15. Despite such implied sophistication, people of African descent who lived on and visited the Comstock during the latter part of the nineteenth century found themselves in a complex political climate that overtly and subtly pervaded many aspects of their lives. For example, in *The Roar and the Silence* (7, 152–153), Ron James tells the story of this group on the Comstock to demonstrate an intriguing pattern of integration, marginal survival, and success. On the one hand, they appeared to have more freedom and opportunity on the Comstock than in many other parts of the country in terms of economic successes and overall integrated living, yet their lives were still not as simple or easy as those of their white neighbors before and after 1881. Racist undertones and overtly restrictive attitudes and laws affected the black population there, and their lives featured a complex juxtaposition of integration and prejudice, neighborly acceptance and ill treatment. Such variation in treatment of African Americans in the West was common, experienced by residents of Virginia City, Nevada, and by African American soldiers stationed throughout the West; see Frank N. Schubert, "Black Soldiers on the White Frontier: Some Factors Influencing Race Relations," *Phylon* 32 (Winter 1971): 411.

16. *Territorial Enterprise,* August 7, 1866.

17. There are no oral histories on record that represent African Americans living in northern Nevada during the recent past; Thomas King, personal communication, 2002.

18. Don McBride, personal communication, 2000.

19. *Stratigraphy* is the description, correlation, and classification of distinct layers of the earth, with each stratum being a homogenous layer that is visually separated from the other layers; see George "Rip" Rapp Jr. and Christopher L. Hill, eds.,

Geoarchaeology: The Earth-Science Approach to Archaeological Interpretation (New Haven, CT: Yale University Press, 1998).

20. Special thanks to Ron James, Tim McCarthy, and Cal Dillon for their assistance with this testing.

21. This was made possible after two years of fundraising resulting in grants from the Nevada State Historic Preservation Office, the National Endowment for the Humanities, and private donors.

22. Also spelled "balk," this term refers to an unexcavated block of earth that archaeologists leave in place between their excavation units. When placed at various intervals throughout an excavation, baulks provide archaeologists with a constant visual reference to the layers they previously exposed.

23. Special thanks to Ahern Rentals of Gardnerville, Nevada, for donating the use of the Bobcat to assist with this next stage of excavation; also thanks to field school student and project member Diane Willis for contacting Ahern Rentals on behalf of the Boston Saloon project. Gratitude to Larry Buhr for being on hand to operate this machinery.

24. In retrospect and now well after many archaeologists have made the technological transition to digital photos, high-quality digital photos should have made up another component of the site's photodocumentation. Because they can be viewed instantly, digital photos allow archaeologists to see if their photos turned out "on the spot." When relying on the more traditional, manual camera photography, unless they had the means to develop negatives in the field, archaeologists could not know whether their field photos would turn out. Since they needed to accurately photograph excavated layers before removing them and excavating further, problems like poor photos or overexposed film could ruin the core record-keeping charter of archaeology for the project in question. The instant feedback available with digital photos allows archaeologists to verify that they have captured the details of the site so that they can move on and excavate the next layer with confidence.

25. Edward Harris, *Principles of Archaeological Stratigraphy,* 2d ed. (London: Academic Press, 1989).

26. Ibid., xiv.

27. The assumption behind this splitting rather than lumping methodology was that it was better to split each stratigraphic context first and then seek similarities among the nature and cultural deposits of each context. Some could then be "lumped" together later; however, if they were lumped together first in the field, potential details of unique context could become lost because of someone's attempt to figure out which general context should be associated with which newly exposed

layer; see also Dixon, *"A Place of Recreation of Our Own,"* appendix A, "Stratigraphic Description of the Boston Saloon Site."

28. Roderick Sprague, "A Functional Classification for Artifacts from Nineteenth- and Twentieth-Century Sites," *North American Archaeologist* 2, no. 3 (1980): 251.

29. Ibid., 259.

30. Donald L. Hardesty et al., *Public Archaeology on the Comstock,* University of Nevada, Reno report prepared for the Nevada State Historic Preservation Office (Carson City: Nevada State Historic Preservation Office, 1996).

31. Dixon et al., "The Archaeology of Piper's Old Corner Bar, Virginia City, Nevada."

32. Dixon, *"A Place of Recreation of Our Own."*

33. Donald L. Hardesty and Ronald M. James, "'Can I Buy You a Drink?': The Archaeology of the Saloon on the Comstock's Big Bonanza" (paper presented at the Mining History Association Conference, Nevada City, California, June 1995); Kelly J. Dixon, "The Archaeology of an Upscale Saloon: Purchasing Champagne, Cigars, and Status at Piper's Old Corner Bar" (paper presented at the Society for Historical Archaeology, Atlanta, 1998); Kelly J. Dixon, "Archaeology of the Boston Saloon: An African American Business in a Western Mining Boomtown" (paper presented at the Society for Historical Archaeology, Long Beach, CA, 2001); Kelly J. Dixon, "The Urban Landscape and Sensory Perception of D Street, Virginia City, Nevada" (paper presented at the 35th annual meeting of the Society for Historical Archaeology, Mobile, AL, 2002); Kelly J. Dixon, "Diversity in the Mining West: DNA and Archaeology at the Boston Saloon" (paper presented at Society for Historical Archaeology, Providence, RI, 2003); Kelly J. Dixon, "From Babylonian Taverns to Western Saloons: Establishing a Temporal Context for Social Drinking" (paper presented at Society for Historical Archaeology, St. Louis, MO, 2004); Kelly J. Dixon, "Survival of Biological Evidence on Artifacts: Applying Forensic Techniques at the Boston Saloon," *Historical Archaeology* 39, no. 1 (forthcoming).

1 ⁓ OPENING SALOON DOORS

1. Ronald M. James, *The Roar and the Silence: A History of Virginia City and the Comstock Lode* (Reno: University of Nevada Press, 1998).

2. Mark Twain [Samuel Clemens, pseud.], *Roughing It* (1873; reprint, New York: Penguin Books, 1981); William Wright [Dan DeQuille, pseud.] *The Big Bonanza* (1876; reprint, Las Vegas: Nevada Publications, 1974).

3. J. Ross Browne, *A Peep at Washoe* (18601861; reprint, Balboa Island, CA: Paisano Press, 1959); J. Ross Browne, "Washoe Revisited," *Harper's Weekly,* 1863.

4. Eliot Lord, *Comstock Mining and Miners* (1883; reprint, Washington, DC: U.S. Geological Survey, Government Printing Office, 1959), 377; Elliott West, *The Saloon on the Rocky Mountain Mining Frontier* (Omaha: University of Nebraska Press, 1979), xiv–xv; Perry Duis, *The Saloon: Public Drinking in Chicago and Boston, 1880–1920* (Urbana: University of Illinois Press, 1983), 1.

5. *Territorial Enterprise,* January–April, 1867, and September 1870; *Virginia Evening Chronicle,* November 1872; *Daily Stage,* September–October 1880.

6. The eight-hour shifts did not represent the standard industrial workday. Rather, Virginia City mines were forced into a twenty-four-hour cycle because the pumps that helped drain water from the underground needed to be operating constantly. The extreme heat at the deeper levels, however, made it physically difficult for miners to put in more than eight hours, and therefore workers in Virginia City had more leisure time than they would have been accustomed to. James, *The Roar and the Silence,* 58, 126, 140–142; see also Mary Martin Murphy, *Mining Cultures: Men, Women, and Leisure in Butte, 1914–1941* (Urbana: University of Illinois Press, 1997).

7. James, *The Roar and the Silence,* 143–166; Susan Lee Johnson, *Roaring Camp: The Social World of the California Gold Rush* (New York: Norton, 2000).

8. West, *The Saloon on the Rocky Mountain Mining Frontier,* 43; Duis, *The Saloon,* 143, 169.

9. Gwendolyn Captain, "Social, Religious, and Leisure Pursuits of Northern California's African American Population: The Discovery of Gold through World War II" (master's thesis, University of California, Berkeley, 1995).

10. Although van Bokkelin was born in New York, his name is of Dutch origin. Misspellings of his name as von Bokkelin have resulted in misunderstandings of his ethnicity as German; see James, *The Roar and the Silence,* 303 n. 38, for a discussion of this debate. Donald L. Hardesty et al., "Public Archaeology on the Comstock," University of Nevada, Reno report prepared for the Nevada State Historic Preservation Office (Carson City: Nevada State Historic Preservation Office, 1996).

11. *Virginia Evening Bulletin,* October 7, 1863; *Territorial Enterprise,* March 6, 1877.

12. *Footlight,* March 13, 1883; *Territorial Enterprise,* March 14, 1883.

13. *Territorial Enterprise,* January 13, 1885.

14. J. Wells Kelly, *J. Wells Kelly's Second Directory of the Nevada Territory, 1863–1864* (Virginia City: Valentine and Company, 1863); Charles Collins, *Mercantile Guide and Directory for Virginia City and Gold Hill, 1864–1865* (Virginia City: Agnew and Deffebach, 1865); *Territorial Enterprise,* August 7, 1866; James, *The Roar and the Silence,* 154.

15. *Territorial Enterprise,* August 7, 1866; *Pacific Appeal,* October 26, 1875; for

general information about African Americans in Virginia City, see Elmer Rusco, *"Good Times Coming?" Black Nevadans in the Nineteenth Century* (Westport, CT: Greenwood, 1975).

16. James, *The Roar and the Silence,* 94, 177.

17. Duis, *The Saloon,* 170.

18. James, *The Roar and the Silence,* 94; Hardesty et al., "Public Archaeology on the Comstock."

19. Mary McNair Mathews, *Ten Years in Nevada: or, Life on the Pacific Coast* (Lincoln: University of Nebraska Press, 1985), 193.

20. West, *The Saloon on the Rocky Mountain Mining Frontier,* 42.

21. Ibid., 53.

22. Lord, *Comstock Mining and Miners,* 93.

23. Donald L. Hardesty and Ronald M. James, "'Can I Buy You a Drink?': The Archaeology of the Saloon on the Comstock's Big Bonanza" (paper presented at the Mining History Association Conference, Nevada City, CA, June 1995).

24. Mark Twain [Samuel Clemens, pseud.], *Roughing It* (1872; reprint, New York: Oxford University Press, 1996), 19.

25. Hamlin Hill, Introduction to *Roughing It,* by Mark Twain (New York: Penguin, 1981), 9.

26. West, *The Saloon on the Rocky Mountain Mining Frontier,* xi, xiii.

27. Ibid., 4, 143.

28. See also Duis, *The Saloon,* and James, *The Roar and the Silence.*

29. See also Hardesty and James, "'Can I Buy You a Drink?'"; Hardesty et al., "Public Archaeology on the Comstock"; Kelly J. Dixon et al., "The Archaeology of Piper's Old Corner Bar, Virginia City, Nevada," Comstock Archaeology Center Preliminary Report of Investigations (Carson City, Nevada: Nevada State Historic Preservation Office, 1999); Kelly J. Dixon, *"A Place of Recreation of Our Own." The Archaeology of the Boston Saloon: Diversity and Leisure in an African American-Owned Saloon, Virginia City, Nevada.* Ann Arbor, MI: University Microfilms International, 2002.

30. See also Adrian Praetzellis and Mary Praetzellis, eds., "Archaeologists as Storytellers," *Historical Archaeology* 32, no. 1 (1998).

31. See also James, *The Roar and the Silence,* 204.

2 ∾ FACADES OF PUBLIC DRINKING

1. In *The Saloon on the Rocky Mountain Mining Frontier* (Lincoln: University of Nebraska Press, 1979), Elliott West describes three phases of boomtown development, with saloon architecture becoming more substantial with each phase; this

architecture changed from tents to ornate structures as mining communities transitioned from camps to cities (36). West also states that because saloons appeared among the first "institutions" in these crude and rudimentary camps, they took on various responsibilities for other institutions that had not yet arrived, such as banks and churches. Early saloons also acted as "institutional newspapers" by communicating information about local news and activities in the mine explorations (76). The price of alcohol also revealed the phase of a community's development, with a drink in a newly founded camp costing twenty-five cents, or two bits. As transportation to a camp improved, supplies became easier to acquire, and more saloon owners appeared. The added competition and increasing availability of alcohol drove the two-bit price down (111). Accordingly, cheaper alcohol signified the existence of an "established" boomtown and the transition from West's so-called first phase to the second and third phases of a mining camp's development.

2. William Wright [Dan DeQuille, pseud.], *The Big Bonanza* (1876; reprint, New York: Knopf, 1953), 36.

3. West, *The Saloon on the Rocky Mountain Mining Frontier*, 36–37.

4. See, for example, *The Good, the Bad, and the Ugly* (Metro-Goldwyn Mayer/United Artists, 1967); *The Wild Bunch* (Warner Brothers, 1969); and *Pat Garrett and Billy the Kid* (Metro-Goldwyn Mayer, 1973).

5. Lloyd Hoff, ed., *The Washoe Giant in San Francisco: Uncollected Sketches by Mark Twain* (San Francisco: George Fields, 1938), 52. Thanks to Ron and Susan James for pointing out this reference.

6. These measurements are estimated from the approximated scale of the bird's-eye view.

7. Donald L. Hardesty et al., *Public Archaeology on the Comstock*, University of Nevada, Reno report prepared for the Nevada State Historic Preservation Office (Carson City: Nevada State Historic Preservation Office, 1996), 57.

8. Ibid.

9. Given the densely built boomtown landscape, a lack of windows along the sides of a building was usually the result of other, one-story buildings' being wedged against that structure; this positioning rendered windows useless at that level. However, the building that contained O'Brien and Costello's establishment was along an alley, which meant there was no neighboring structure to explain the lack of windows in this case.

10. Hardesty et al., "Public Archaeology on the Comstock," 28.

11. See, for example, *The Good, the Bad, and the Ugly.*

12. See ibid.

13. *Virginia Evening Chronicle*, July 11, 1876, 3.

14. Dixon et al., "The Archaeology of Piper's Old Corner Bar," 181–183.

3 ⌢ AUTHENTIC SALOON SETS

1. See also Kelly J. Dixon, "The Urban Landscape and Sensory Perception of D Street, Virginia City, Nevada" (paper presented at the 35th annual meeting of the Society for Historical Archaeology, Mobile, AL, 2002).

2. Augustus Koch, "Bird's-eye View of Virginia City, Nevada" (San Francisco: Britton, Rey, and Company, 1875).

3. Hardesty et al., "Public Archaeology on the Comstock," University of Nevada, Reno report prepared for the Nevada State Historic Preservation Office (Carson City: Nevada State Historic Preservation Office, 1996), 58; R. Scott Baxter, "Beer, Bourbon, and Bullets: Munitions Analysis from a Nineteenth-Century Bar and Shooting Gallery" (paper presented at the 31st annual meeting of the Society for Historical Archaeology, Atlanta, 1998).

4. Eliot Lord, *Comstock Mining and Miners* (1883; reprint, Washington, DC: U.S. Geological Survey, Government Printing Office, 1959), 73; also cited in Ronald M. James, *The Roar and the Silence* (Reno and Las Vegas: University of Nevada Press, 1998), 186.

5. Hardesty et al., "Public Archaeology on the Comstock," 28–29.

6. www.armstrong.com/reslinoleumna/linoleum_history.jsp.

7. U.S. Patent Records, Patent No. 134,281, December 24, 1872; microfilm at Getchell Library, University of Nevada, Reno.

8. See, for example, Elliott West, *The Saloon on the Rocky Mountain Mining Frontier* (Lincoln: University of Nebraska Press, 1979), 42.

9. Kelly J. Dixon et al., "The Archaeology of Piper's Old Corner Bar, Virginia City, Nevada," Comstock Archaeology Center Preliminary Report of Investigations (Carson City: Nevada State Historic Preservation Office, 1999), 184, 186.

10. www.tinceiling.com/background/tinhistory.htm.

11. Stephen J. Gould, "Seeing Eye to Eye," *Natural History* (July–August) 1997: 16–19, 60–62. Thanks to Rick Fields for suggesting this reference.

12. *Carson Daily Appeal*, June 13, 1873, 2.

13. Donald L. Hardesty and Ronald M. James, "'Can I Buy You a Drink?': The Archaeology of the Saloon on the Comstock's Big Bonanza" (paper presented at the Mining History Association Conference, Nevada City, CA, June 1995), 8.

14. For the purposes of this book, the term *identity* is used to imply a collectivity with others that is realized by understanding oneself as participating within a culture; it is not necessarily fixed, however; it may change and may be dependent on context. See Bruce A. Knauft, *Genealogies for the Present in Cultural Anthropology* (New

York: Routledge, 1996), 235–236; Maria Franklin, "The Archaeological Dimensions of Soul Food: Interpreting Race, Culture, and Afro-Virginian Identity," in *Race and the Archaeology of Identity,* edited by Charles E. Orser Jr., 90 (Salt Lake City: University of Utah Press, 2001).

15. Kenneth L. Brown, "Interwoven Traditions: Archaeology of the Conjurer's Cabins and the African American Cemetery at the Levi Jordan and Frogmore Plantations" (paper presented at the "Places of Cultural Memory: African Reflections on the American Landscape" conference, Atlanta, 2001); Elizabeth J. Himelfarb, "Hoodoo Cache," *Archaeology* 53, no. 3 (2000): 21.

16. Brown, "Interwoven Traditions"; James M. Davidson, "Rituals Captured in Context and Time: Charm Use in North Dallas Freedmen's Town (1869–1907), Dallas, Texas," *Historical Archaeology* 38, no. 2 (2004): 22–54.

17. Patrick E. Martin, personal communications, 2003.

18. Davidson, "Rituals Captured in Context and Time," 25–27; James M. Davidson, e-mail correspondence, 2004.

19. Davidson, "Rituals Captured in Context and Time," 28–38.

20. Brown, "Interwoven Traditions"; James M. Davidson, "Haunts, Poison, Sickness, Conjure: An Exploration of Conjuration and Counter Charms in the African-American South, 1869–1915" (paper presented at the 35th annual meeting of the Society for Historical Archaeology, Mobile, AL), 12.

21. Davidson, "Rituals Captured in Context and Time," 40.

22. Brown, "Interwoven Traditions," 108.

23. Ronald DuPont Jr., "Of Things Left Inside Walls," American Cultural Resources Association, public listserv; acra-1@lists.nonprofit.net, 2003; June Swann, "Shoes Concealed in Buildings," *Costume Society Journal* 30 (1996): 65–66.

24. Julie M. Schablitsky, "The Other Side of the Tracks: The Archaeology and History of a Virginia City, Nevada Neighborhood" (Ph.D. diss., Portland State University, 2002), 223–224.

25. Jessica Axom and Ron James, personal communication, August 4, 2004.

26. DuPont, "Of Things Left Inside Walls."

27. Ralph Merrifield, *The Archaeology of Ritual and Magic* (New York: New Amsterdam Books, 1987); Swann, "Shoes Concealed in Buildings," 56–69; Schablitsky, "The Other Side of the Tracks," 223.

28. James, *The Roar and the Silence,* 198; see also Duane A. Smith, "Comstock Miseries: Medicine and Mining in the 1860s," *Nevada Historical Society Quarterly* 36, no. 1 (Spring 1993): 9.

29. Tommy-knockers did not usually reveal themselves visibly to miners but were often heard knocking on rock walls and wood beams; their tapping sounds were said

to warn miners of possible accidents or imminent collapses. If their warnings were ignored, they became mischievous and were blamed for cave-ins and the loss of tools and lunches. For a detailed treatment of this topic, see Ronald M. James, "Knockers, Knackers, and Ghosts: Immigrant Folklore in the Western Mines," *Western Folklore* 51 (1992) 2: 153–177.

4 ⁓ MENU ITEMS

1. Advertisements for saloon lunches in Virginia City were quite common, such as those for the Enterprise Saloon, the Express Saloon, and the International Saloon, published in the *Territorial Enterprise,* January 1, 1867, 4, and the *Virginia Evening Chronicle,* November 4, 1872, 1; see also Joseph R. Conlin, *Bacon, Beans, and Gallantines: Food and Foodways on the Western Mining Frontier* (Reno and Las Vegas: University of Nevada Press, 1986); Conlin notes, for example: "It would appear that the provision of food for customers was at least a sideline of most saloons . . . [and] the saloon-restaurant combination was a fixture in the mining camps" (174, 176).

2. G. Dodd (1851), www.outlawcook.com/Page0116.html; the original work has not been found.

3. *Virginia Evening Chronicle,* November 4, 1872, 1.

4. *Territorial Enterprise*, March 9, 1867; *Virginia Evening Chronicle,* November 4, 1872, 1.

5. For example, green bottles and green bottle fragments make up 25 percent of the Boston Saloon collection, 26 percent of the total Piper's Old Corner Bar collection. They make up 20 to 22 percent of the Hibernia collection; this percentage is the result of tallying all catalog entries for glass items described as "olive green, "dark olive green," or "green kick-up" within the Hibernia's stratigraphic context. The exact percentage of the total O'Brien and Costello's collection is unknown at this time. However, green glass is quite prevalent as one goes through the various boxes containing that collection.

6. Donald L. Hardesty and Ronald M. James, "'Can I Buy You a Drink?': The Archaeology of the Saloon on the Comstock's Big Bonanza" (paper presented at the Mining History Association Conference, Nevada City, CA, June 1995), 6; Kelly J. Dixon et al., "The Archaeology of Piper's Old Corner Bar, Virginia City, Nevada," Comstock Archaeology Center Preliminary Report of Investigations (Carson City: Nevada State Historic Preservation Office, 1999).

7. www.ginvodka.org; according to the gin and vodka distillers' Web site, the best-known premium gin brands are British in origin. Since this statement comes from the industry's Web site, it should be used with caution since it likely represents a form of self-promotion. Even so, the fact that the Gordon's Gin bottle fragments at the

Boston Saloon came from London implies that that establishment used one of the premium brands of British origin, the presence of which represented yet another upscale component contributing to the ambience of that saloon.

8. Lynn Blumenstein, *Wishbook 1865: Relic Identification for the Year 1865* (Salem, OR: Old Time Bottle Publishing Company, 1968), 87–89.

9. Dixon et al., "The Archaeology of Piper's Old Corner Bar," 61.

10. Eliot Lord, *Comstock Mining and Miners* (1883; reprint, Washington, DC: U.S. Geological Survey, Government Printing Office, 1959), 93.

11. Ibid.

12. Henry Bradley, *A New English Dictionary on Historical Principles Founded Mainly on the Materials Collected by the Philological Society,* vol. 4, edited by James A. H. Murray (Oxford: Clarendon, 1901), 169.

13. *Territorial Enterprise,* November 24, 1866.

14. Arthur Kallet and F. J. Schlink, *100,000,000 Guinea Pigs: Dangers in Everyday Foods, Drugs, and Cosmetics* (New York: Vanguard, 1933), 151.

15. Ronald R. Switzer, *The Bertrand Bottles: A Study of Nineteenth-Century Glass and Ceramic Containers* (Washington, DC: National Park Service, U.S. Department of the Interior, 1974), 76; Rex Wilson, *Bottles on the Western Frontier* (Tucson: University of Arizona Press, 1981), 35.

16. Donald L. Hardesty and Eugene M. Hattori, "Archaeological Studies in the Cortez Mining District," Technical Report no. 8 (Reno: Bureau of Land Management, 1981), 39.

17. Hardesty and James, "Can I Buy You a Drink?," 7.

18. For example, Morton's Sample Rooms, located at 119 South C Street in Virginia City, advertised bitters and essences with their alcohol products; *Territorial Enterprise,* February 7, 1867, 1.

19. Switzer, *The Bertrand Bottles,* 76; Wilson, *Bottles on the Western Frontier,* 35.

20. For soda water ads associated with Virginia City stores, refer to the *Territorial Enterprise,* October 22, 1870.

21. Hardesty and James ("Can I Buy You a Drink?" 7) note that some of the soda and mineral water bottles recovered from the Hibernia Brewery represent Irish-owned companies, including the Eagle Soda Works in Sacramento, California, and the Cantrell and Cochrane Company of Dublin and Belfast; for information on the Cantrell and Cochrane Company and its history, see www.cantrellandcochrane.com.

22. Peter D. Schulz, Betty J. Rivers, Mark M. Hales, Charles A. Litzinger, and Elizabeth A. McKee, *The Bottles of Old Sacramento: A Study of Nineteenth-Century Glass and Ceramic Retail Containers,* part 1. California Archaeological Reports, no. 20 (Sacramento: State of California Department of Parks and Recreation, Cultural

Resources Management Unit, 1980), 114. Thanks to Carrie Smith for suggesting this reference.

23. Robert Leavitt, personal communication, 2003; for more details on stoneware mineral water jugs, see Robert Leavitt, "Taking the Waters: Stoneware Jugs and the Taste of Home They Contained" (master's thesis, University of Nevada, Reno, Department of Anthropology, 2004).

24. Robert Leavitt, "Appendix B. Stoneware 'Selters' Bottle Analysis," in Dixon, "The Archaeology of Piper's Old Corner Bar, Virginia City, Nevada."

25. Robert Leavitt, personal communication, 2002. Archaeologist Robert Leavitt conducted an analysis of the various positions of these bottles and bottle fragments using Geographic Information System (GIS) technology to conclude that the bottles were stored above a cellar but fell into that space during a devastating fire in the Piper's building in 1883.

26. Robert Leavitt, "Gin Jugs and Mineralwasser: A Taste of Home" (paper presented at the 35th annual meeting of the Society for Historical Archaeology, Mobile, AL, 2002).

27. Ibid.

28. Schulz et al., "The Bottles of Old Sacramento," 111; Robert Leavitt, personal communication, 2003.

29. Ronald M. James, *The Roar and the Silence* (Reno and Las Vegas: University of Nevada Press, 1998), 59.

30. This flume conveyed potable water from Marlette Lake in the Sierra Nevada to Virginia City using an inverted siphon; for details, see Hugh A. Schamberger, *Historic Mining Camps of Nevada, Water Supply for the Comstock: Early History, Development, Water Supply* (prepared in cooperation with Nevada Department of Conservation and Natural Resources and U.S. Geological Survey, 1969).

31. See also Dixon et al., "The Archaeology of Piper's Old Corner Bar," and Dan L. Urriola, "Appendix D: Mended Ceramic Artifact Assemblage," in ibid.

32. Thanks to Dr. Stephanie Livingston for her assistance with the Boston Saloon and Piper's Old Corner Bar bone analysis. Identifications of the species and element of bone samples from the Boston Saloon and Piper's Old Corner Bar made use of the private osteological collections of Dr. Livingston, a portion of the Nevada State Museum's collection, and a sheep specimen from the University of Nevada's School of Agriculture's Meat Technology studies. Each bone analyzed required a description of provenience, taxon, element, side, portion, meat cut, illustrations (when applicable), age, butchering evidence, burn damage, and weathering. Elizabeth Scott conducted a faunal analysis for the Hibernia Brewery and O'Brien and Costello's estab-

lishment; see Elizabeth M. Scott, "Faunal Remains From the Hibernia Saloon and Faunal Remains From the O'Brien and Costello Bar and Shooting Gallery," in *Public Archaeology on the Comstock,* by Donald L. Hardesty et al., 64–88, University of Nevada, Reno report prepared for the Nevada State Historic Preservation Office (Carson City: Nevada State Historic Preservation Office, 1996).

33. Ibid., 85. Upon comparing the faunal data from the Hibernia Brewery and O'Brien and Costello's Saloon and Shooting Gallery with the faunal remains from two other saloons, including Cronin's Oyster Saloon in Sacramento and a tavern in rural Wisconsin, Scott concludes that all of the types of establishments have a similar faunal signature, "whether rural or urban, Western or Midwestern, working class or upper class, and regardless of the ethnicity of the owners, cooks, or clientele . . . the assemblages are dominated by domestic mammals, chickens, turkeys, ducks, and geese"; Thomas A. Wake, "Appendix K: Faunal Report, Zooarchaeology of the Pantheon Saloon and Its Local Area, Skagway, Alaska," in *Archaeological Excavations in Skagway,* vol. 9, *Excavations at the Pantheon Saloon Complex,* by Tim A. Kartdatzke, K–4 (Anchorage: National Park Service, 2002).

34. Scott, "Faunal Remains From the Hibernia Saloon and Faunal Remains From the O'Brien and Costello Bar and Shooting Gallery," 85.

35. Dixon, *"A Place of Recreation of Our Own." The Archaeology of the Boston Saloon: Diversity and Leisure in an African American-Owned Saloon, Virginia City, Nevada* (Ann Arbor, MI: University Microfilms International, 2002), 147–158. Working with zooarchaeologist Stephanie Livingston, Dixon found that sheep is the most abundant taxon represented among the Boston Saloon's animal bones, making up about 34 percent of the analyzed materials. This is followed by unidentified mammal remains, which make up 23 percent of the analyzed specimens. The unidentified category is closely followed by cow, which makes up 19 percent of the analyzed materials. Rodent is next, at 15 percent of the analyzed material. Pig makes up only about 1.3 percent of the analyzed sample, with other animals such as duck, chicken, and rabbit combined making up only about 2 percent of the sample.

36. Because of the nonidentifiable bones in the Piper's sample, only 47 of the specimens display enough of the bone to confidently assess and identify the taxa represented. Of those, only a small number have enough bone to assess body parts and associated meat cuts.

37. Scott, "Faunal Remains From the Hibernia Saloon and Faunal Remains From the O'Brien and Costello Bar and Shooting Gallery," 80–86; Jessica Leigh Kinchloe, "'The Best the Market Affords': Food Consumption at the Merchant's Exchange Hotel, Aurora, Nevada" (master's thesis, University of Nevada, Reno,

2001), 64, 73–74, 114, 117; Julie Schablitsky, "The Other Side of the Tracks: The Archaeology and History of a Virginia City, Nevada, Neighborhood" (Ph.D. diss., Portland State University, 2002).

38. Other faunal research made it possible to make assertions about the quality and cost associated with animal bones at the Boston Saloon; see Peter D. Schulz and Sherri M. Gust, "Faunal Remains and Social Status in Nineteenth-Century Sacramento," *Historical Archaeology* 17, no. 1 (1983): 44–53. The respective values of meat cuts were determined using nineteenth-century retail values extrapolated from Schulz and Gust (48) and using Kinchloe, "'The Best the Market Affords,'" 78 and 85.

39. Dixon, *"A Place of Recreation of Our Own,"* 147–158. The sheep, cow, and pig specimens bring the total items from the Boston Saloon that reflect high-quality cuts of meat to 343 (61 percent). The medium-quality cuts of sheep, cow, and pig across the site totaled 65 (12 percent), and the low-quality cuts from these animals add up to 104 (19 percent); 45 (8 percent) specimens represent unidentifiable meat cuts.

40. Diana C. Crader, "Slave Diet at Monticello," *American Antiquity* 55, no. 4 (1990): 690–717; Crader provides a diagram showing the higher and lower quality of meat associated with pig remains from the colonial South; the diagram indicates that the higher-quality pig cuts came from the loin area, front legs, and back legs, representing pork chops and hams (699).

41. G. Dodd (1851), www.outlawcook.com/Page0116.html.

42. Scott, "Faunal Remains From the Hibernia Saloon and Faunal Remains From the O'Brien and Costello Bar and Shooting Gallery," 86.

43. Ibid., 80–81.

44. Ibid., 80.

45. Dixon, *"A Place of Recreation of Our Own,"* 171; excavations at the Boston Saloon, for example, yielded among its massive faunal collection only one Pacific oyster, one eastern oyster, and fragments of oysters and clams; native, or rock, oysters came from the Pacific Coast from southern California to southeast Alaska, while eastern oysters came from the Atlantic Coast; thanks to Julie Schablitsky for her assistance with this.

46. Conlin, *Bacon, Beans, and Gallantines,* 120.

47. Ronald M. James, personal communication, 2003.

48. Richard Erdoes, *Saloons of the Old West* (New York: Knopf, 1979), 109–114; see also G. Dodd, www.outlawcook.com/Page0116.html.

49. Shane Bernard and Ashley Dumas, personal communication, 2002. Tabasco Sauce company historian Shane Bernard and archaeology student Ashley Dumas noted this bottle's rarity. The bottle's angular shoulder and its embossed basal mark, with the words "TABASCO//PEPPER//SAUCE" and with embossed six-pointed stars,

were not rare, but the Boston Saloon's bottle stood out because it had those traits in combination with a relatively thin lip. Up to that point, Tabasco historians believed that the earliest bottles made especially for the pepper sauce had much thicker lips.

50. Charles E. Orser and David W. Babson, "Tabasco Brand Pepper Sauce Bottles from Avery Island, Louisiana," *Historical Archaeology* 25 (1990) 2: 107.

51. Shane Bernard and Ashley Dumas, personal communication, 2002.

52. This cannot be proved, however, and is an example of the ways in which artifacts can lead archaeologists only so far before their interpretations become mere speculation; Ntozake Shange, *If I Can Cook/You Know God Can* (Boston: Beacon, 1998), 29.

53. Schablitsky, "The Other Side of the Tracks."

54. Wilson, *Bottles on the Western Frontier,* 82.

55. Thanks to Stephanie Livingston and Donald Grayson for their help in identifying these bones; special thanks to Robert Kopperl for transporting the bones.

5 ⁓ A TOAST TO THE ARTIFACTS

1. For example, *Maverick* (Warner Brothers, 1995).

2. See for example, the *Bonanza* episode titled "The Gunmen" (National Broadcasting Company, 1959/1973); *The Good, the Bad, and the Ugly* (Metro-Goldwyn Mayer/United Artists, 1967); *Once Upon a Time in the West* (Paramount Studios, 1968); *Pat Garrett and Billy the Kid* (Metro-Goldwyn Mayer, 1973); *Tombstone* (Hollywood Pictures, 1993).

3. See, for example, George L. Miller, "A Revised Set of CC Index Values for Classification and Economic Scaling of English Ceramics From 1787 to 1880," *Historical Archaeology* 25 (1993) 1:1–25; see also www.thepotteries.org/timeline/index.htm

4. For more information on the significance of linking local archaeological sites with a global system, see Donald L. Hardesty, "Evolution on the Industrial Frontier," in *The Archaeology of Frontiers and Boundaries,* edited by Stanton W. Green and Stephen M. Perlman, 213–229 (New York: Academic Press, 1983); Andre Gunder Frank and Barry K. Gills, eds., *The World System: Five Hundred Years or Five Thousand?* (London: Routledge, 1996); Charles Orser Jr., *A Historical Archaeology of the Modern World* (New York: Plenum, 1996); Gil J. Stein, *Rethinking World-Systems: Diasporas, Colonies, and Interaction in Uruk Mesopotamia* (Tucson: University of Arizona Press, 1999).

5. Ralph Kovel and Terry Kovel, *Kovels' New Dictionary of Marks, Pottery and Porcelain: 1850 to the Present* (New York: Crown, 1986).

6. The process of extracting chemical information from this stain is addressed in chapter 7, which is devoted to forensic techniques.

7. Elizabeth M. Scott, "Faunal Remains From the Hibernia Saloon and Faunal Remains From the O'Brien and Costello Bar and Shooting Gallery," in *Public Archaeology on the Comstock,* by Donald L. Hardesty et al., 80–81, University of Nevada, Reno report prepared for the Nevada State Historic Preservation Office (Carson City: Nevada State Historic Preservation Office, 1996); Scott notes that many of the bones from the Hibernia Brewery came in the form of "finger foods" such as sheep and pig's feet.

8. Donald L. Hardesty and Ronald M. James, "'Can I Buy You a Drink?': The Archaeology of the Saloon on the Comstock's Big Bonanza" (paper presented at the Mining History Association Conference, Nevada City, CA, June 1995), 8.

9. Dan L. Urriola, "Appendix D: Mended Ceramic Artifact Assemblage," in "The Archaeology of Piper's Old Corner Bar, Virginia City, Nevada," by Kelly J. Dixon et al., Comstock Archaeology Center Preliminary Report of Investigations (Carson City: Nevada State Historic Preservation Office, 2001).

10. Donald L. Hardesty et al., *Public Archaeology on the Comstock,* University of Nevada, Reno report prepared for the Nevada State Historic Preservation Office (Carson City: Nevada State Historic Preservation Office, 1996), 39.

11. Ronald M. James, *The Roar and the Silence* (Reno and Las Vegas: University of Nevada Press, 1998), 204.

12. *Territorial Enterprise,* March 27, 1867, 4.

6 ⁓ DESIRES FOR DIVERSION

1. Mark Twain [Samuel Clemens, pseud.], *Mark Twain in Virginia City, Nevada* (Las Vegas: Nevada Publications, 1985), 69–71.

2. Ibid., 71.

3. For other vices, see William Wright [Dan DeQuille, pseud.], *The Big Bonanza* (1876; reprint, New York: Knopf, 1953), 295–296, and Julie M. Schablitsky, "The Magic Wand: Hypodermic Drug Injection of the Nineteenth Century" (paper presented at the 35th annual meeting of the Society for Historical Archaeology, Mobile, AL, 2002). As noted by Wright and Schablitsky, opium smoking and morphine injection were among the other vices in boomtowns such as Virginia City; however, these vices were not necessarily affiliated with saloons to the same degree that smoking and gambling were.

4. Michael A. Pfeiffer, "The Tobacco-Related Artifact Assemblage from the Martinez Adobe, Pinole, California," in *The Archaeology of the Clay Tobacco Pipe,* part 7, *America,* edited by Peter Davey, 185–194 (Oxford, England: B.A.R. Series 175, 1983).

5. The following archaeologists visited the Boston Saloon excavation as consult-

ants and were asked whether they had ever encountered the red clay pipes during their experiences with nineteenth-century urban and/or African American sites: Paul Mullins 2000, personal communication; Adrian and Mary Praetzellis 2000, personal communication; questions were also posed about the red clay pipe to archaeologist and tobacco pipe expert Michael "Smoke" Pfeiffer in e-mail correspondence (2002) and to international tobacco pipe expert David Higgins in e-mail correspondence (2003). Thanks to Troy Lawrence for conducting a final, last-minute search on this pipe style.

6. Twain, *Mark Twain in Virginia City, Nevada,* 70.

7. *Territorial Enterprise,* August 7, 1866; *Territorial Enterprise*, March 9, 1867; *Territorial Enterprise,* March 27, 1867; Eliot Lord, *Comstock Mining and Miners* (1883; reprint, Washington, DC: U.S. Geological Survey, Government Printing Office, 1959), 73; Ronald M. James, *The Roar and the Silence* (Reno and Las Vegas: University of Nevada Press, 1998), 186; see also John M. Findlay, *People of Chance: Gambling in American Society From Jamestown to Las Vegas* (New York: Oxford University Press, 1986).

8. The poker chips came out of a stratigraphic layer that contains materials that both coincide with and postdate the occupation of the Boston Saloon. This means that the poker chips may or may not be associated with the operation of that business; however, the game of poker is noted as being played in this establishment, according to a *Territorial Enterprise* article (August 7, 1866), so the chances are good that the chips are indeed affiliated with the saloon's occupation of the corner of D and Union Streets.

9. www.tradgames.org.uk.

10. Donald L. Hardesty and Ronald M. James, "'Can I Buy You a Drink?': The Archaeology of the Saloon on the Comstock's Big Bonanza" (paper presented at the Mining History Association Conference, Nevada City, CA, June 1995), 8.

11. www.ariga.com/southernjourney/saturday.htm.

12. James, *The Roar and the Silence,* 211, 260; James points out that even though prostitution was a "cornerstone of Comstock society," many community members had rather mixed feelings about it, which means that even though some people perceived prostitutes as entrepreneurs and prostitution as a legitimate way of making a living, others certainly saw it as a disrespectable occupation; thus prostitution was also perceived as a vice.

13. See also Elliott West, *The Saloon on the Rocky Mountain Mining Frontier* (Omaha: University of Nebraska Press, 1979), 48; James, *The Roar and the Silence,* 185–186; and Hardesty and James, "Can I Buy You a Drink?"

14. Stephen A. Mrozowski, Grace H. Ziesing, and Mary C. Beaudry, *Living on*

the Boott: Historical Archaeology at the Boott Mills Boardinghouses, Lowell, Massachusetts (Amherst: University of Massachusetts Press, 1996), 79–80; special thanks to Jonathan Hardes for lending me this book at exactly the right time.

15. James, *The Roar and the Silence,* 185–186.

16. Mary McNair Methews, *Ten Years in Nevada: or, Life on the Pacific Coast* (Lincoln: University of Nebraska Press, 1985), 195; see also Jan I. Loverin and Robert A. Nylen, "Creating a Fashionable Society: Comstock Needleworkers From 1860 to 1880," in *Comstock Women: The Making of a Mining Community,* edited by Ronald M. James and C. Elizabeth Raymond (Reno and Las Vegas: University of Nevada Press, 1998), 115–116.

17. Mathews, *Ten Years in Nevada,* 193.

18. James, *The Roar and the Silence,* 179–180.

19. Ibid., 178.

20. Hardesty et al., "Public Archaeology on the Comstock," 25; James, *The Roar and the Silence,* 204.

21. For more information on calicoes as "everyday" dresses, see Peggy Ann Osborne, *About Buttons: A Collector's Guide 150* A.D. *to the Present* (Lancaster, PA: Schiffer, 1993).

22. *Territorial Enterprise,* March 9, 1867; *Footlight,* February 20, 1878.

23. James, *The Roar and the Silence,* 176; Sue Fawn Chung, "Their Changing World: Chinese Women on the Comstock," in *Comstock Women: The Making of a Mining Community,* edited by Ronald M. James and C. Elizabeth Raymond, 204–228 (Reno: University of Nevada Press, 1998).

24. Mathews, *Ten Years in Nevada,* 165.

25. Myron Angel, ed., *History of Nevada* (Oakland, CA: Thomson and West, 1881), 188; Elmer Rusco, *"Good Times Coming?" Black Nevadans in the Nineteenth Century* (Westport, CT: Greenwood, 1975), 56; James, *The Roar and the Silence,* 154, 300 n. 25; the woman noted as William Brown's wife testified in court after her husband shot and killed an African American named John Scott; the records of the event indicate that John Scott was obsessed with attacking William Brown in the latter's saloon, for unknown reasons. Under duress from Scott's threats, Brown repeatedly asked Scott to go away and at some point shot him. The authorities likely concluded that the shooting was self-defense, and William Brown was presumably not convicted of any crime, since he appears again in historical records as a citizen in good standing only two years later. Given the ambiguous nature of this case in historical records, there may have been an underlying but unspoken notion in the white community that there was little need to prosecute a black for a black crime; see James, *The Roar and the Silence,* 154).

26. Linda Krugner, e-mail correspondence, 2002; Jan Loverin, personal communication, 1999.

27. *Pacific Coast Business Directory for 1867* (San Francisco: Henry G. Langley, Langley Publishing, 1871).

28. See James, *The Roar and the Silence,* 154, for a more thorough overview of Amanda Payne.

29. Thomas D. Phillips, "The Black Regulars," in *The West of the American People,* edited by Allan G. Bogue, Thomas D. Phillips, and James E. Wright, 138–143 (Itasca, IL: Peacock, 1970); Kenneth W. Porter, "Black Cowboys in the American West, 1866–1900," in *African Americans on the Western Frontier,* edited by Monroe Lee Billington and Roger D. Hardaway (Niwot: University Press of Colorado, 1998), 110–127. Racially segregated leisure prevailed in the nineteenth-century American West, especially where white women were concerned. For example, Porter observes informal segregation within the same saloon, with white cowboys served at one end of the saloon and black cowboys at the other. While blacks and whites came into contact in certain informally segregated gambling halls and saloons, brothel owners maintained more strict lines of segregation because the majority of prostitutes were white. However, as Kenneth Porter points out, some "soiled doves of color" worked in the cattle towns, providing a segregated alternative for black cowboys barred from brothels staffed with white prostitutes; "Black Cowboys in the American West," 124. Additionally, the U.S. Army barred black men from dancing with white women. Phillips notes that on occasion black soldiers attended dances given by white soldiers, with an underlying understanding that the black soldiers would not seek dance partners among the white women; "The Black Regulars," 140. See also Charles E. Orser Jr., ed., *Race and the Archaeology of Identity* (Salt Lake City: University of Utah Press, 2001).

30. Cathy Spude, e-mail correspondence, 2003.

31. John Taylor Waldorf, *A Kid on the Comstock: Reminiscences of a Virginia City Childhood* (Reno: University of Nevada Press, 1970), 62.

32. Lord, *Comstock Mining and Miners,* 73; also cited in James, *The Roar and the Silence,* 186.

33. For example, films such as *Tombstone* (Hollywood Pictures, 1993) and *Maverick* (Warner Brothers, 1995) featured scuffles—or potential scuffles—resulting from gambling disagreements. Other well-known Hollywood productions that have perpetuated the association with brawls and saloons include *The Good, the Bad, and the Ugly* (Metro-Goldwyn Mayer, 1967) and *Bonanza* episodes such as "The Gunman" (National Broadcasting Company, 1973).

7 ~ CRIME SCENE INVESTIGATION?

1. Forensic studies were made possible through the collaboration of many individuals: Julie M. Schablitsky provided inspiration and brainstorming to carry out the research; a grant from the Nevada State Historic Preservation Office provided funding to complete the lab work necessary to prepare artifacts for forensic analyses; a grant from the University of Nevada, Reno Graduate Student Association supplied funding for the DNA and GCMS testing; Raymond Grimsbo and staff at Intermountain Forensic Laboratories ran the DNA and GCMS tests to fit a tight budget; and finally, Stephanie Livingston gave the author her introduction to forensic anthropology and generously gave in-kind zooarchaeological consultation with the Boston Saloon's bone collection.

2. The traditional medico-legal boundary for forensic anthropology focuses upon people who died within the last fifty years, while archaeology tends to focus on people who died fifty-plus years ago.

3. Chi-square (χ^2) tests were conducted to evaluate the relative abundance of, and relationships between, burned bones at the Boston Saloon site. In general, chi-square tests provide a statistical means of organizing observed versus expected data. These tests were used to make viable comparisons of certain observed data, such as frequencies of burned and/or calcined bone, with an expected distribution of data, such as the presence of more burned specimens in the dump area. The relative abundance of different frequencies between the burned specimens from the two areas at the Boston Saloon site was shown by chi square test results: ($\chi^2 = 16.32$; df = 1; $\pm = 0.05$); similarly, the relative abundance of bones with calcine damage between the two areas was also revealed: $\chi^2 = 17.99$; df = 2; $\pm = 0.05$. The abbreviation df refers to degrees of freedom, which is the total number of tests [in a chi square analysis] minus one. The $\pm = 0.05$ refers to a commonly used significance (alpha) level on a chi-square probability distribution table.

4. James Deetz, *In Small Things Forgotten: An Archaeology of Early American Life* (New York: Anchor Books/Doubleday, 1996), 254–255.

5. Thomas Loy, "Prehistoric Blood Residues: Detection on Tool Surfaces and Identification of Species of Origin," *Science* 220 (1983): 1269–1271; Thomas Loy and A. R. Wood, "Blood Residue Analysis of Cayonu Tepesi, Turkey," *Journal of Field Archaeology* 16 (1989): 451–460; Thomas Loy and B. L. Hardy, "Blood Residue Analysis of 90,000-Year-Old Stone Tools from Tabun Cave, Israel," *Antiquity* 66 (1992): 24–35; Thomas Loy and E. James Dixon, "Blood Residues on Fluted Points From Eastern Alaska," *American Antiquity* 63 (1998) 1: 21–46; Martin Jones, *The Molecule Hunt: Archaeology and the Search for Ancient DNA* (New York: Arcade, 2001).

6. Volunteers Dan Urriola, Chris Kruse, and Kinsey Kruse called attention to the stain while Urriola was in the process of mending shards to reconstruct the stoneware crocks to near-complete vessels.

7. Jack Nowicki, "Analysis of Chemical Protection Sprays by Gas Chromatography/Mass Spectroscopy," *Journal of Forensic Sciences* 27 (1982) 3: 704–709.

8. Jones, *The Molecule Hunt,* 10, 14.

9. Bernd Herrmann and Susanne Hummel, eds., *Ancient DNA: Recovery and Analysis of Genetic Material from Paleontological, Archaeological, Museum, Medical, and Forensic Specimens* (New York: Springer-Verlag, 1994), 1–2; Jones, *The Molecule Hunt,* 36–37.

10. UV light causes DNA molecules to become cross-linked; see also Norah McNally et al., "Evaluation of Deoxyribonucleic Acid (DNA) Isolated From Human Bloodstains Exposed to Ultraviolet Light, Heat, Humidity, and Soil Contamination," *Journal of Forensic Sciences* 34 (1989) 5: 1062.

11. Although the presence of only one individual's DNA implied that archaeologists did not contaminate the pipe stem during recovery or cataloging, the DNA profile was quite intact. As a matter of fact, it represents the most intact known DNA profile ever recovered from an inanimate object in an archaeological context. The sample's relatively undamaged quality cast doubt on its authenticity as aDNA. To address this problem, the forensic lab conducted further tests to verify whether the profile actually represented an uncontaminated profile or whether a member of our archaeology crew was responsible for leaving such an intact DNA signature on this object. For example, a female member of the crew who handled the pipe stem submitted samples of her own DNA to the forensic lab. Her DNA profile differed from the original profile, indicating that the sample had not been contaminated by the archaeology crew. Control tests on three other objects from the Boston Saloon tested negative for DNA, indicating that human exogenous DNA did not douse the site's artifact assemblage. Such controls emphasize the unique quality of the DNA profile on the pipe stem and reinforce the validity of that profile's uncontaminated nature. It is therefore quite likely that the DNA actually did represent that of a woman from the nineteenth century, considering the pipe stem's provenience in nineteenth-century stratigraphic deposits.

12. Ancestral background may be determined by the distribution of alleles and allele variants on a DNA profile; see Bruce Budowle, Tamyra R. Moretti, Anne L. Baumstark, Debra A. Defenbaugh, and Kathleen M. Keys, "Population Data on the Thirteen CODIS Core Short Tandem Repeat Loci in African Americans, U.S. Caucasians, Hispanics, Bahamians, Jamaicans, and Trinidadians," *Journal of Forensic Sciences* 44 (1994) 6: 1277–1286; see also Bruce Budowle, Brendan Shea, Stephen

Niezgoda, and Ranajit Chakraborty, "CODIS STR Loci Data from 41 Sample Populations," *Journal of Forensic Sciences* 46 (2001) 3: 453–489; in her research with aDNA on a syringe needle, Julie Schablitsky found a number of DNA profiles, with at least one exhibiting a rare allele variant indicating that a man of African descent received a hypodermic injection from that needle; see Julie M. Schablitsky, "The Magic Wand: Hypodermic Drug Injection of the Nineteenth Century" (paper presented at the 35th annual meeting of the Society for Historical Archaeology, Mobile, AL, 2002).

13. For information on the ways in which black men and white women "did not mix" in leisure contexts, see Thomas D. Phillips, "The Black Regulars," in *The West of the American People,* edited by Allan G. Bogue, Thomas D. Phillips, and James E. Wright, 140 (Itasca, IL: Peacock, 1970); see also William L. Lang, "Helena, Montana's Black Community, 1900–1912," in *African Americans on the Western Frontier,* edited by Monroe Lee Billington and Roger D. Hardaway, 202 (Niwot: University Press of Colorado, 1998).

14. For more information on the syntheses of documentary and material evidence, see Lu Ann De Cunzo and Bernard L. Herman, eds., *Historical Archaeology and the Study of American Culture* (Knoxville: University of Tennessee Press, 1996), 6.

15. See, for example, Elliott West, *The Saloon on the Rocky Mountain Mining Frontier* (Lincoln: University of Nebraska Press, 1979), 145.

16. While some nineteenth-century writers refer to tobacco use as a vice (e.g., Mark Twain, [Samuel Clemens, pseud.], *Mark Twain in Virginia City, Nevada* [Las Vegas: Nevada Publications, 1985], 69–71), smoking still did not have the negative associations then that it does today; Stephen A. Mrozowski, Grace H. Ziesling, and Mary C. Beaudry, *Living on the Boott: Historical Archaeology at the Boott Mills Boardinghouses, Lowell, Massachusetts* (Amherst: University of Massachusetts Press, 1996), 67–68; and Lauren J. Cook's "Tobacco-Related Material Culture and the Construction of Working Class Culture," in *Interdisciplinary Investigation of the Boott Mills, Lowell, Massachusetts,* vol. 3, *The Boardinghouse System as a Way of Life,* edited by Mary C. Beaudry and Stephen A. Mrozowski (Boston: U.S. Government Printing Office for the North Atlantic Region (Boston) of the National Park Service, 1989), make note of this and also briefly address women and tobacco use during the nineteenth century. In regard to the latter they argue that smoking at that time was primarily limited to men. Women did participate in smoking, but they risked the stigma of being considered loose if they did so.

17. Julie Schablitsky, "Genetic Archaeology: The Recovery and Interpretation of Nuclear DNA From a Nineteenth-Century Hypodermic Syringe," *Historical Archaeology,* forthcoming.

18. *Columbia Encyclopedia,* 6th ed. (New York: Columbia University Press, 2001).

19. Arthur Kallet and F. J. Schlink, *100,000,000 Guinea Pigs: Dangers in Everyday Foods, Drugs, and Cosmetics* (New York: Vanguard, 1933); see also www.drugs.uta.edu.

20. For example, in the movie *Tombstone* (1993), Wyatt Earp's wife, Mattie, was portrayed as a laudanum addict.

21. While women are shown smoking cigarettes and cigarillos in films such as *Hanny Caulder* (1971), there is only a single instance, in a relatively popular film, *The Outlaw Josey Wales* (Malpaso/Warner Brothers, 1971), of a woman shown smoking a tobacco pipe; thanks to Cris Borgnine and Richard P. Benjamin for assisting with the finer points of women and tobacco use in the western film genre.

22. Julie Schablitsky, personal communication, 2003.

23. For a discussion of historical archaeology as a "humanistic science," see Charles E. Orser Jr. and Brian M. Fagan, *Historical Archaeology* (New York: Harper Collins College Publishers, 1995), 191–195, 214–248.

CONCLUSION ~ CASTING THE SALOON OF THE WILD WEST IN A NEW LIGHT

1. Wes D. Gehring, *Handbook of American Film Genres* (London: Greenwood Press, 1988), 26; Elliott West, *The Saloon on the Rocky Mountain Mining Frontier* (Lincoln: University of Nebraska Press, 1979), xi, xii, 143; in his research on Rocky Mountain saloons, regional cultural historian Elliott West credited films, television, and novels with introducing the public to the significance of the saloon in the history of the West.

2. Hortense Powdermaker, *Hollywood, the Dream Factory: An Anthropologist Looks at the Movie-makers* (Boston: Little, Brown, 1950); thanks to G. G. Weix for pointing out this reference.

3. D. Dean, *Museum Exhibition: Theory and Practice* (London: Routledge, 1994), 30; thanks to Richard P. Benjamin for pointing out this reference.

4. Mark Twain [Samuel Clemens, pseud.], *Mark Twain in Virginia City, Nevada* (Las Vegas: Nevada Publications, 1985), 112; the origins of writers sensationalizing gunfighters and violence during the nineteenth century are accurately underscored in the Clint Eastwood film *The Unforgiven* (Warner Brothers, 1992).

5. R. Scott Baxter, "Beer, Bourbon, and Bullets: Munitions Analysis from a Nineteenth-Century Bar and Shooting Gallery" (paper presented at the 31st annual meeting of the Society for Historical Archaeology, Atlanta, 1998). Today such recreational use of guns may be seen as a form of violent leisure; however, Baxter underscores the

status-based pastime of using and owning guns for European immigrants in the American West (10); such activities were reserved for aristocrats in the Old World and perhaps empowered the new immigrants with heightened social status.

6. See, for example, *Once Upon a Time in the West* (Paramount Studios, 1968); *The Good, the Bad, and the Ugly* (Metro-Goldwyn Mayer/United Artists, 1967); *The Wild Bunch* (Warner Brothers, 1969); *Pat Garrett and Billy the Kid* (Metro-Goldwyn Mayer, 1973); and *Tombstone* (Hollywood Pictures, 1993); thanks to Richard P. Benjamin for assisting with the research related to these films.

7. Mark Twain [Samuel Clemens, pseud.], *Roughing It* (1873; reprint, New York: Penguin, 1981), 303.

8. *Territorial Enterprise,* August 8, 1866.

9. Maurice Halbwachs, *On Collective Memory* (Chicago: University of Chicago Press, 1992); thanks to Elizabeth Harvey for introducing me to this reference.

10. William Hayes Ward, *The Seal Cylinders of Western Asia* (Washington, DC: Carnegie Institution of Washington, 1910), 1–7.

11. Briggs Buchanan, *Early Near Eastern Seals in the Yale Babylonian Collection* (New Haven, CT: Yale University Press, 1981), ix.

12. D. J. Wiseman, *Cylinder Seals of Western Asia* (London: Batchworth, 1958), 25.

13. H. Frankfort, *Cylinder Seals: A Documentary Essay on the Art and Religion of the Ancient Near East* (London: Macmillan, 1939), 77–78; thanks to Professor A. Millard and to Richard Paul Benjamin at the University of Liverpool for making the search for this and other related references more efficient. The first documentation of social drinking from the Near East came from seals used in the ancient Mesopotamian city of Sumer. Sometimes the seals show a single figure, such as a goddess, drinking from a vessel placed on the floor using a straw or a tube; while many of the banquet scenes were ritual-based and linked with certain gods or goddesses, Sumerians clearly had many occasions for secular festivities, as shown by carvings on certain seals of human figures. These included figures such as a man and a woman—often shown facing each other on either side of a large jar, or drinking vessel—sharing beer using tube or straws.

14. Patrons frequently toasted the dead, in ancient Egyptian drinking houses; Roger Protz, *Ultimate Encyclopedia of Beer: The Complete Guide to the World's Greatest Brews* (London: Carlton Books, 1995); see also www.fosters.com.au/beer/history/history_of_beer.asp.

15. Frederick W. Hackwood, *Inns, Ales, and Drinking Customs of Old England* (New York: Sturgis and Walton, 1909), 30; see also www.ancientroute.com/cities/Pelusium.htm; also, the Egyptian Book of the Dead described a beer brewed from

barley, and this was the primary menu item at drinking houses; E. A. Wallis Budge, *The Book of the Dead: Facsimile of the Papyrus of Ani in the British Museum,* 2d ed. (London: Harrison and Sons, 1894).

16. Protz, *Ultimate Encyclopedia of Beer*; although the Romans had a history of wine making, many taverns served beer using grains that were easier to grow than grapes throughout many parts of Europe.

17. Hackwood, *Inns, Ales, and Drinking Customs of Old England,* 31–32.

18. West, *The Saloon on the Rocky Mountain Mining Frontier,* 26.

19. Ibid.

20. David Waldstreicher, *In the Midst of Perpetual Fetes: The Making of American Nationalism, 1776–1820* (Chapel Hill: University of North Carolina Press, 1997), 26.

21. West, *The Saloon on the Rocky Mountain Mining Frontier,* 27.

22. Ibid.; Perry Duis, *The Saloon: Public Drinking in Chicago and Boston, 1880–1920* (Urbana: University of Illinois Press, 1983), 5–6, 10.

23. West, *The Saloon on the Rocky Mountain Mining Frontier,* 27.

24. Donald L. Hardesty et al., *Public Archaeology on the Comstock,* University of Nevada, Reno report prepared for the Nevada State Historic Preservation Office (Carson City: Nevada State Historic Preservation Office, 1996), 5.

25. Joel L. Swerdlow, "Changing America," *National Geographic* 200, no. 3 (2001): 42–61. The celebration of distinct cultures and diversity actually resists the "melting pot" assumption of assimilation into one massive culture. For background on diversity as a principal issue of anthropology, archaeology's parent discipline, see Bruce A. Knauft, *Genealogies for the Present in Cultural Anthropology* (New York: Routledge, 1996), 25. For information on why archaeology can be used by living people to better understand contemporary human issues, see Charles E. Orser Jr., *A Historical Archaeology of the Modern World* (New York: Plenum, 1996), 199.

26. Stephen Steinberg, *The Ethnic Myth: Race, Ethnicity, and Class in America* (Boston: Beacon, 2001), 5.

27. West, *The Saloon on the Rocky Mountain Mining Frontier,* 43; see also Duis, *The Saloon,* 143, 169.

28. See for example, Frederik Barth, ed., *Ethnic Groups and Boundaries: The Social Organization of Culture Difference* (Boston: Little, Brown, 1967).

29. Ronald M. James, "Defining the Group: Nineteenth-Century Cornish on the Mining Frontier," in *Cornish Studies: 2*, edited by Philip Payton (Exeter: University of Exeter Press, 1994), 32–47; Gwendolyn Captain, "Social, Religious, and Leisure Pursuits of Northern California's African American Population: The Discovery of Gold through World War II" (master's thesis, University of California, Berkeley, 1995).

30. Leisure studies call attention to the fact that people tend to express their ethnic or cultural, class-based, gender-based identity during their free time, especially when living in a prejudicial social and economic context. By treating saloons as a leisure institution, it is possible to examine them as places for people to express their identities. See, for example, Oscar Handlin, *Boston's Immigrants, 1790–1865: A Study in Acculturation* (Cambridge: Harvard University Press, 1941); West, *The Saloon on the Rocky Mountain Mining Frontier*; Gunther Barth, *City People: The Rise of Modern City Culture in Nineteenth-Century America* (New York: Oxford University Press, 1980); Hugh Cunningham, *Leisure and the Industrial Revolution c. 1780–1880* (New York: St. Martin's, 1980); Duis, *The Saloon*; Captain, "Social, Religious, and Leisure Pursuits of Northern California's African American Population"; Mary Martin Murphy, *Mining Cultures: Men, Women, and Leisure in Butte, 1914–1941* (Urbana: University of Illinois Press, 1997).

31. West, *The Saloon on the Rocky Mountain Mining Frontier*, 91.

32. *Footlight*, February 20, 1878.

33. Elmer Rusco, *"Good Times Coming?" Black Nevadans in the Nineteenth Century* (Westport, CT: Greenwood, 1975); Eugene M. Hattori, *Northern Paiutes on the Comstock: Archaeology and Ethnohistory of an American Indian Population in Virginia City, Nevada*, Occasional Papers, no. 2, Nevada State Museum, edited by Donald R. Tuohy and Doris L. Rendall (Carson City: Nevada State Museum, 1975); Ronald M. James, *The Roar and the Silence* (Reno and Las Vegas: University of Nevada Press, 1998), 36; Ronald M. James and C. Elizabeth Raymond, eds., *Comstock Women: The Making of a Mining Community* (Reno: University of Nevada Press, 1998); Sue Fawn Chung, "Their Changing World: Chinese Women on the Comstock," in *Comstock Women*, 204–228,; Eugene M. Hattori, "'And Some of Them Swear like Pirates': Acculturation of American Women in Nineteenth-Century Virginia City," in *Comstock Women*, 229–245.

34. *Territorial Enterprise*, August 7, 1866; U.S. Manuscript Census for 1870 on Microfilm; *Virginia and Truckee Railroad Directory 1873–1874*; *Pacific Appeal*, October 16, 1876.

35. Captain, "Social, Religious, and Leisure Pursuits of Northern California's African American Population," 55; Randall B. Woods, "Integration, Exclusion, or Segregation? The 'Color Line' in Kansas, 1878–1900," in *African Americans on the Western Frontier*, edited by Monroe Lee Billington and Roger D. Hardaway, 128 (Niwot: University Press of Colorado, 1998).

36. Quintard Taylor, *In Search of the Racial Frontier: African Americans in the American West, 1528–1990* (New York: Norton, 1998), 1–17; see also Rusco, *"Good Times Coming?"*

37. Captain, "Social, Religious, and Leisure Pursuits of Northern California's African American Population," 42–44; Roger D. Hardaway, "The African American Frontier: A Bibliographic Essay," in *African Americans on the Western Frontier,* 240; see also Taylor, *In Search of the Racial Frontier.*

38. U.S. Manuscript Census on Microfilm [and available in computer database format at the Nevada State Historic Preservation Office] 1860, 1870, 1880; Nevada State Census on Microfilm 1863, 1875; J. Wells Kelly, *J. Wells Kelly's Second Directory of Nevada Territory, 1863–1864* (Virginia City: Valentine and Company, 1863); Charles Collins, *Charles Collins Mercantile Guide and Directory for Virginia City and Gold Hill, 1864–1865* (Virginia City: Agnew and Deffebach, 1865); John F. Uhlhorn, *Virginia and Truckee Railroad Directory, 1873–1874;* the quote "a place of recreation of our own" came from the *Pacific Appeal,* October 16, 1875. Stories of prejudicial treatment were documented by writers such as Thomas Detter in the following references: *Pacific Appeal,* October 8, 1870; *Pacific Appeal,* November 26, 1870; Farah J. Griffin and Cheryl J. Fisk, *A Stranger in the Village: Two Centuries of African-American Travel Writing* (Boston: Beacon, 1998), 30–31; see also *Pacific Appeal,* May 1868; *Pacific Appeal,* April 11, 1868; *Pacific Appeal,* December 31, 1870.

39. Eric Foner, *Reconstruction: America's Unfinished Revolution, 1863–1877* (New York: Harper and Row, 1988).

40. Duis, *The Saloon,* 160.

41. For Example, Paul Mullins, *Race and Affluence: An Archaeology of African America and Consumer Culture* (New York: Kluwer Academic/Plenum Publishers, 1999); Adrian Praetzellis and Mary Praetzellis, "We Were There, Too": Archaeology of an African-American Family in Sacramento, California, Cultural Resources Facility, Anthropological Studies Center (Rohnert Park, CA: Sonoma State University, 1992); Adrian Praetzellis and Mary Praetzellis, "Mangling Symbols of Gentility in the Wild West," *American Anthropologist* 103, no. 3 (2001): 651.

42. *Territorial Enterprise,* April 5, 1870; Rusco, *"Good Times Coming?"* 78; James, *The Roar and the Silence,* 153.

43. Brown's success is certainly evident in historical records. For example, the Boston Saloon operated for at least one or two years during the early 1860s at its first location, on B Street, and another nine years (1866–1875) at the D Street location. This indirectly suggests that the Boston Saloon flourished. It also indicates that William Brown was a successful entrepreneur. He was able to keep his establishment in operation for many years despite the difficulty of operating a saloon throughout the unpredictable boom and bust cycles of a mining community and despite the challenges of being an African American business owner in a city with relatively few

African Americans. Alternatively, the paucity of African Americans in this boomtown made whites more likely to accept blacks because they lacked sufficient numbers to pose a threat; see Taylor, *In Search of the Racial Frontier,* 17.

44. For example, Noel Ignatiev, *How the Irish Became White* (New York: Routledge, 1995).

45. James, *The Roar and the Silence,* 178–179.

46. Ignatiev, *How the Irish Became White*; Ronald M. James, "Defining the Group: Nineteenth-Century Cornish on the Mining Frontier," in *Cornish Studies: 2,* edited by Philip Payton, 32 (Exeter: University of Exeter Press, 1994).

47. James, "Defining the Group," 32.

48. Ignatiev, *How the Irish Became White.*

49. Ibid.

50. Virginia City's "Big Four" who became enormously wealthy were John Mackay, O'Brien, Fair, and Flood; James, *The Roar and the Silence,* 102–105.

51. Ignatiev, *How the Irish Became White*; see also David M. Emmons, *The Butte Irish: Class and Ethnicity in an American Mining Town, 1875–1925* (Urbana: University of Illinois Press, 1989); Glenna Matthews, "Forging a Cosmopolitan Civic Culture: The Regional Identity of San Francisco and Northern California," in *Many Wests: Place, Culture, and Regional Identity,* edited by David M. Wrobel and Michael C. Steiner, 219 (Lawrence: University of Kansas Press, 1997).

52. See also Mullins, *Race and Affluence,* viii, 32; Praetzellis and Praetzellis, "Mangling Symbols of Gentility in the Wild West," 651.

53. *Territorial Enterprise,* April 5, 1870; Rusco, *"Good Times Coming?"* 16, 23; James, *The Roar and the Silence,* 152–153.

54. Todd Guenther, "At Home on the Range: Black Settlement in Rural Wyoming, 1850–1950" (master's thesis, University of Wyoming, 1988), 133; Frank N. Schubert, "Black Soldiers on the White Frontier: Some Factors Influencing Race Relations," *Phylon* 32 (Winter 1971): 411.

55. The documentary record adds another dimension to the sounds of brass instruments associated with African Americans in Virginia City. On September 19, 1877, the *Territorial Enterprise* notes festivities in honor of the dedication of a new hall for the community's Black Masonic Lodge, that is, the Prince Hall Masons' Ashlar Lodge. This ceremony included a substantial meal and dancing to the music of a black band with a white leader. The link between such a band and the mouthpiece at the Boston Saloon is impossible to prove at this point, but it is entirely possible that some of the band's performers entertained the Boston Saloon's customers.

56. By the 1980s many archaeologists had begun to address studies of gender as part of a movement that critiqued their colleagues for promoting erroneous repre-

sentations of women—or for overlooking them altogether—in archaeological work. This inspired alternative interpretations of the past that attempted to address issues such as gender roles, identity, and ideology; see Margaret W. Conkey and Janet Spector, "Archaeology and the Study of Gender," *Advances in Archaeological Method and Theory* 7 (1984): 15; Joan Gero and Margaret W. Conkey, eds., *Engendering Archaeology: Women and Prehistory* (Oxford: Blackwell, 1991); Donald L. Hardesty, "Gender and Archaeology on the Comstock," 283–302, in *Comstock Women*; Margaret Purser, "Gender Archaeology," in *Archaeological Method and Theory: An Encyclopedia,* edited by Linda Ellis, 233–237 (New York: Garland, 2000).

57. This differentiation could be the result of misinterpretation of feminine versus masculine artifacts.

58. Susan Lee Johnson, *Roaring Camp: The Social World of the California Gold Rush* (New York: Norton, 2000), 157, 187, 276, 286; Johnson presents an egalitarian "frontier" setting that was somewhat tolerant of people from various backgrounds and of various reputations, calling attention to the ways in which the arrival of "white Anglo women" caused a redirection of the more decadent mining camp leisure activities such as heavy drinking in saloons, dancing at fandangos, and sexual activities in brothels toward balls and parties sponsored by community-based organizations. Johnson asserts that that setting began to disappear by the 1850s when "Anglo control" over the southern California gold rush mines took shape as class rule. Thus the previous social world of "possibility and permeable boundaries began to give way to more entrenched forms of dominance rooted in Anglo American constructions of gender, of class position, and of race, ethnicity, and nation—that is, in particular ideas about what it meant to live like 'white folks'"; for a discussion of the development of the middle-class value system, which can be traced back to the early nineteenth century, see William L. Barney, *The Passage of the Early Republic: An Interdisciplinary History of Nineteenth-Century America* (Lexington, MA: D. C. Heath, 1987), 77–82; Diana diZerega Wall, *The Archaeology of Gender: Separating the Spheres in Urban America* (New York: Plenum, 1994); see also Nan A. Rothschild, *New York City Neighborhoods: The Eighteenth Century* (New York: Academic Press, 1990), 183.

59. Murphy, *Mining Cultures,* 43; although Murphy underscores the fact that social drinking became divided by gender, she discusses how this changed during Prohibition, as women in Butte, Montana, were heavily involved in making "moonshine."

60. See for example, James, *The Roar and the Silence,* 185–186.

61. Murphy, *Mining Cultures,* 43.

62. Captain, "Social, Religious, and Leisure Pursuits of Northern California's African American Population," 93–94; Captain argues that African Americans used

leisure activities such as benevolent societies, fraternal orders, reading groups, and churches as a means of self-improvement.

63. *Pacific Coast Business Directory, 1871–1873*; Nevada State Census on Microfilm 1875; James, *The Roar and the Silence*, 154.

64. This situation should cause archaeologists to consider that every seemingly "genderless" tobacco pipe they unearth may have a fair chance of representing women.

65. See also Anders Andrén, *Between Artifacts and Texts: Historical Archaeology in Global Perspective* (New York: Plenum, 1997), 179–182.

66. Sir Herbert Butterfield, *The Whig Interpretation of History* (London: G. Bell and Sons, 1950); thanks to Ron James for pointing out this reference.

67. Kathleen Deagan, "Historical Archaeology's Contributions to Our Understanding of Early America," in *Historical Archaeology in Global Perspective*, edited by Lisa Falk, 108–109 (Washington, DC: Smithsonian Institution Press, 1991).

68. Emmanuel Le Roy Ladurie, *Montaillou: The Promised Land of Error* (New York: Vintage Books, a Division of Random House, 1979), 278. Historian Emmanuel Le Roy Ladurie examined documents related to an inquisition of peasants in the village of Montaillou, southern France, conducted by Bishop Jacques Fournier between 1318 and 1325. Fournier's inquisition involved interviews with peasants in this region, providing direct testimony from those individuals and insights into their lives and worldviews. Among these insights are the peasants' concepts of time, and what can be interpreted as "leisure time." Le Roy Ladurie addressed concepts of time and space, noting how the people of Montaillou were devoted to their various forms of work—farming, herding, and crafts. However, he also observed that their workdays were "punctuated with long, irregular pauses, during which one would chat with a friend, perhaps at the same time enjoying a glass of wine." From such observations, it is apparent that peasant life in southern France during the early fourteenth century integrated so-called leisure activities, such as visiting with a friend over a glass of wine, with everyday work activities. This concept of time or "free time" in pre-capitalist Europe eventually disintegrated as capitalism took over as a way of life; thanks to Ron James for introducing me to this reference.

69. For more on historical archaeology's ability to "unromanticize" the past, see James Deetz, *In Small Things Forgotten: An Archaeology of Early American Life* (New York: Anchor Books/Doubleday, 1996), 254–255.

70. For a discussion of historical archaeology's ability to shed light on minute and particular details, see Charles E. Orser Jr. and Brian M. Fagan, *Historical Archaeology* (New York: Harper Collins College Publishers, 1995), 19.

71. See ibid., 20.

72. Molefi Kete Asante, *The Afrocentric Idea,* rev. and expanded ed. (Philadelphia: Temple University Press, 1998), xi.

73. Andren, *Between Artifacts and Texts*; Larry Lankton, personal communication, 1994.

74. The Bonanza is inspired by the television series *Bonanza,* and Julie Bulette's is named in honor of a darling prostitute who was murdered in 1867; see also Susan A. James, "Queen of Tarts," *Nevada Magazine* 44, no. 5 (September–October 1984): 51–54.

BIBLIOGRAPHY

BOOKS AND ARTICLES

Andrén, Anders. *Between Artifacts and Texts: Historical Archaeology in Global Perspective.* New York: Plenum, 1998.

Angel, Myron, ed. *History of Nevada.* Oakland, CA: Thomson and West, 1881.

Asante, Molefi Kete. *The Afrocentric Idea.* Rev. and expanded ed. Philadelphia: Temple University Press, 1998.

Barney, William L. *The Passage of the Early Republic: An Interdisciplinary History of Nineteenth-Century America.* Lexington, MA: D. C. Heath, 1987.

Barth, Frederik, ed. *Ethnic Groups and Boundaries: The Social Organization of Culture Difference.* Reprint. Boston: Little, Brown, 1967.

Barth, Gunther. *City People: The Rise of Modern City Culture in Nineteenth-Century America.* New York: Oxford University Press, 1980.

Baxter, R. Scott. "Beer, Bourbon, and Bullets: Munitions Analysis From a Nineteenth-Century Bar and Shooting Gallery." Paper presented at the 31st annual meeting of the Society for Historical Archaeology, Atlanta, 1998.

Blumenstein, Lynn. *Wishbook 1865: Relic Identification for the Year 1865.* Salem, OR: Old Time Bottle Publishing Company, 1968.

Bradley, Henry. *A New English Dictionary on Historical Principles Founded Mainly on the Materials Collected by the Philological Society.* Vol. 4. Edited by James A. H. Murray. Oxford: Clarendon Press, 1901.

Brown, Grafton T. "Bird's-eye View of Virginia City." San Francisco: Britton and Company, 1861.

Brown, Kenneth L. "Interwoven Traditions: Archaeology of the Conjurer's Cabins and the African American Cemetery at the Levi Jordan and Frogmore Plantations." Paper presented at the "Places of Cultural Memory: African Reflections on the American Landscape" conference, Atlanta, 2001.

———. "Material Culture and Community Structure: The Slave and Tenant Community at Levi Jordan's Plantation, 1848–1892." In *Working Toward Freedom,* edited by Larry E. Hudson, Jr., 95–118. Rochester, NY: University of Rochester Press, 1994.

Brown, Robert L. *Saloons of the American West: An Illustrated Chronicle.* Denver: Robert L. Brown, 1978.

Browne, J. Ross. *A Peep at Washoe.* 1860–1861. Reprint, Balboa Island, CA: Paisano Press, 1959.

———. "Washoe Revisited." *Harper's Weekly,* 1863.

Buchanan, Briggs. *Early Near Eastern Seals in the Yale Babylonian Collection.* New Haven, CT: Yale University Press, 1981.

Budge, E. A. Wallis. *The Book of the Dead: Facsimile of the Papyrus of Ani in the British Museum.* 2d ed. London: Harrison and Sons, 1894.

Budowle, Bruce, Tamyra R. Moretti, Anne L. Baumstark, Debra A. Defenbaugh, and Kathleen M. Keys. "Population Data on the Thirteen CODIS Core Short Tandem Repeat Loci in African Americans, U.S. Caucasians, Hispanics, Bahamians, Jamaicans, and Trinidadians." *Journal of Forensic Sciences* 44, no. 6 (1999): 1277–1286.

Budowle, Bruce, Brendan Shea, Stephen Niezgoda, and Ranajit Chakraborty. "CODIS STR Loci Data from 41 Sample Populations." *Journal of Forensic Sciences* 46, no. 3 (2000): 453–489.

Butterfield, Sir Herbert. *The Whig Interpretation of History.* London: G. Bell and Sons, 1950.

Captain, Gwendolyn. "Social, Religious, and Leisure Pursuits of Northern California's African American Population: The Discovery of Gold Through World War II." Master's thesis, University of California, Berkeley, 1995.

Chung, Sue Fawn. "Their Changing World: Chinese Women on the Comstock." In *Comstock Women: The Making of a Mining Community,* edited by Ronald M. James and C. Elizabeth Raymond, 204–228. Reno and Las Vegas: University of Nevada Press, 1998.

Conkey, Margaret W., and Janet Spector. "Archaeology and the Study of Gender." *Advances in Archaeological Method and Theory* 7 (1984): 1–38.

Conlin, Joseph R. *Bacon, Beans, and Gallantines: Food and Foodways on the Western Mining Frontier.* Reno and Las Vegas: University of Nevada Press, 1986.

Cook, Lauren J. "Tobacco-Related Material Culture and the Construction of Working Class Culture." In *Interdisciplinary Investigation of the Boott Mills, Lowell, Massachusetts,* Vol. 3, *The Boardinghouse System as a Way of Life,* edited by

Mary C. Beaudry and Stephen A. Mrozowski. Boston: U.S. Government Printing Office for the North Atlantic Region (Boston) of the National Park Service, 1989.

Crader, Diana C. "Slave Diet at Monticello." *American Antiquity* 55, no. 4 (1990): 690–717.

Cunningham, Hugh. *Leisure and the Industrial Revolution, c. 1780–1880.* New York: St. Martin's, 1980.

Davidson, James M. "Haunts, Poison, Sickness, Conjure: An Exploration of Conjuration and Counter Charms in the African-American South, 1869–1915." Paper presented at the 35th annual meeting of the Society for Historical Archaeology, Mobile, Alabama.

———. "Rituals Captured in Context and Time: Charm Use in North Dallas Freedmen's Town (1869–1907), Dallas, Texas." *Historical Archaeology* 38, no. 2 (2004): 22–54.

Deagan, Kathleen. "Historical Archaeology's Contributions to Our Understanding of Early America." In *Historical Archaeology in Global Perspective,* edited by Lisa Falk, 97–112. Washington, DC: Smithsonian Institution Press, 1991.

Dean, D. *Museum Exploration: Theory and Practice.* London: Routledge, 1994.

De Cunzo, Lu Ann, and Bernard L. Herman, eds. *Historical Archaeology and the Study of American Culture.* Knoxville: University of Tennessee Press, 1996.

Deetz, James. Foreword to *A Chesapeake Family and Their Slaves: A Study in Historical Archaeology,* by Anne Elizabeth Yentsch, xviii–xx. Cambridge: Cambridge University Press, 1994.

———. *In Small Things Forgotten: An Archaeology of Early American Life.* New York: Anchor Books/Doubleday, 1996.

Dixon, Kelly J. "Archaeology of the Boston Saloon: An African American Business in a Western Mining Boomtown." Paper presented at the Society for Historical Archaeology, Long Beach, CA, 2001.

———. "The Archaeology of Piper's Old Corner Bar, Virginia City, Nevada," With contributions by Ronald M. James, Robert C. Leavitt, Dan Urriola, and Chris Urriola. Comstock Archaeology Center Preliminary Report of Investigations. Carson City: Nevada State Historic Preservation Office, 1999.

———. "The Archaeology of an Upscale Saloon: Purchasing Champagne, Cigars, and Status at Piper's Old Corner Bar." Paper presented at the Society for Historical Archaeology, Atlanta, 1998.

———. "Diversity in the Mining West: DNA and Archaeology at the Boston Saloon." Paper presented at the Society for Historical Archaeology, Providence, RI, 2003.

———. "From Babylonian Taverns to Western Saloons: Establishing a Temporal

Context for Social Drinking." Paper presented at Society for Historical Archaeology, St. Louis, MO, January 2004.

———. *"A Place of Recreation of Our Own." The Archaeology of the Boston Saloon: Diversity and Leisure in an African American-Owned Saloon, Virginia City, Nevada.* Ann Arbor, MI: University Microfilms International, 2002.

———. "Survival of Biological Evidence on Artifacts: Applying Forensic Techniques at the Boston Saloon." *Historical Archaeology* 39, no. 1 (forthcoming).

———. "The Urban Landscape and Sensory Perception of D Street, Virginia City, Nevada." Paper presented at the 35th annual meeting of the Society for Historical Archaeology, Mobile, AL, 2002.

Dodd, G. Cited at www.outlawcook.com/Page0116.html; the original work has yet to be found. 1851.

Duis, Perry. *The Saloon: Public Drinking in Chicago and Boston, 1880–1920.* Urbana: University of Illinois Press, 1983.

Dunnell, Robert. "The Notion Site." In *Space, Time, and Archaeological Landscapes,* edited by Jacqueline Rossignol and LuAnn Wandsnider, 21–41. New York: Plenum, 1992.

DuPont, Ronald, Jr. "Of Things Left Inside Walls." American Cultural Resources Association public listserv; acra–1@lists.nonprofit.net, 2003.

Ebert, James. *Distributional Archaeology.* Albuquerque: University of New Mexico Press, 1992.

Emmons, David M. *The Butte Irish: Class and Ethnicity in an American Mining Town, 1875–1925.* Urbana: University of Illinois Press, 1989.

Erdoes, Richard. *Saloons of the Old West.* New York: Knopf, 1979.

Findlay, John M. *People of Chance: Gambling in American Society From Jamestown to Las Vegas.* New York: Oxford University Press, 1986.

Fliess, Kenneth H. "There's Gold in Them Thar—Documents? The Demographic Evolution of Nevada's Comstock, 1860 Through 1910, and the Intersection of Census Demography and Historical Archaeology." *Historical Archaeology* 34, no. 2 (2000): 65–88.

Foner, Eric. *Reconstruction: America's Unfinished Revolution, 1863–1877.* New York: Harper and Row, 1988.

Frank, Andre Gunder, and Barry K. Gills, eds. *The World System: Five Hundred Years or Five Thousand?* London and New York: Routledge, 1996.

Frankfort, H. *Cylinder Seals: A Documentary Essay on the Art and Religion of the Ancient Near East.* London: Macmillan, 1939.

Franklin, Maria. "The Archaeological Dimensions of Soul Food: Interpreting Race, Culture, and Afro-Virginian Identity." In *Race and the Archaeology of Identity,* edited by Charles E. Orser Jr., 88–107. Salt Lake City: University of Utah Press, 2001.

Gehring, Wes D. *Handbook of American Film Genres.* London: Greenwood, 1988.

Gero, Joan, and Margaret W. Conkey, eds. *Engendering Archaeology: Women and Prehistory.* Oxford: Blackwell, 1991.

Gould, Stephen J. "Seeing Eye to Eye." *Natural History* (July–August 1997): 16–19, 60–62.

Griffin, Farah J., and Cheryl J. Fisk. *A Stranger in the Village: Two Centuries of African-American Travel Writing.* Boston: Beacon, 1998.

Guenther, Todd. "At Home on the Range: Black Settlement in Rural Wyoming, 1850–1950." Master's thesis, University of Wyoming, 1988.

Hackwood, Frederick W. *Inns, Ales, and Drinking Customs of Old England.* New York: Sturgis and Walton, 1909.

Halbwachs, Maurice. *On Collective Memory.* Chicago: University of Chicago Press, 1992.

Handlin, Oscar. *Boston's Immigrants, 1790–1865: A Study in Acculturation.* Cambridge, MA: Harvard University Press, 1941.

Hardaway, Roger D. "The African American Frontier: A Bibliographic Essay." In *African Americans on the Western Frontier,* edited by Monroe Lee Billington and Roger D. Hardaway, 231–257. Niwot: University Press of Colorado, 1998.

Hardesty, Donald L. "Evolution on the Industrial Frontier." In *The Archaeology of Frontiers and Boundaries,* edited by Stanton W. Green and Stephen M. Perlman, 213–229. New York: Academic Press, 1985.

Hardesty, Donald L., with Jane E. Baxter, Ronald M. James, Ralph B. Giles Jr., and Elizabeth M. Scott. "Public Archaeology on the Comstock." University of Nevada, Reno report prepared for the Nevada State Historic Preservation Office. Carson City: Nevada State Historic Preservation Office, 1996.

Hardesty, Donald L., and Eugene M. Hattori. "Archaeological Studies in the Cortez Mining District, 1981." Technical Report no. 8. Reno: Bureau of Land Management, 1982.

Hardesty, Donald L., and Ronald M. James. "'Can I Buy You a Drink?': The Archaeology of the Saloon on the Comstock's Big Bonanza." Paper presented at the Mining History Association Conference, Nevada City, CA, June 1995.

Harris, Edward. *Principles of Archaeological Stratigraphy.* 2d ed. London: Academic Press, 1989.

Hattori, Eugene M. *Northern Paiutes on the Comstock: Archaeology and Ethnohistory of an American Indian Population in Virginia City, Nevada.* Carson City: Nevada State Museum, 1975.

———. "'And Some of Them Swear like Pirates': Acculturation of American Women in Nineteenth-Century Virginia City." In *Comstock Women: The Making of a Mining Community*, edited by Ronald M. James and C. Elizabeth Raymond, 229–245. Reno and Las Vegas: University of Nevada Press, 1998.

Herrmann, Bernd, and Susanne Hummel, eds. *Ancient DNA: Recovery and Analysis of Genetic Material From Paleontological, Archaeological, Museum, Medical, and Forensic Specimens.* New York: Springer-Verlag, 1994.

Hill, Hamlin. Introduction to *Roughing It*, by Mark Twain. New York: Penguin, 1981.

Himelfarb, Elizabeth J. "Hoodoo Cache." *Archaeology* 53, no. 3 (2000).

Hoff, Lloyd, ed. *The Washoe Giant in San Francisco: Uncollected Sketches by Mark Twain.* San Francisco: George Fields, 1938.

Ignatiev, Noel. *How the Irish Became White.* New York: Routledge, 1995.

James, Ronald M. "African Americans on the Comstock: A New Look." Paper presented at the Conference on Nevada History, Reno, 1997.

———. "Defining the Group: Nineteenth-Century Cornish on the Mining Frontier." In *Cornish Studies:* 2, edited by Philip Payton, 32–47. Exeter: University of Exeter Press, 1994.

———. "Erin's Daughters on the Comstock: Building Community." In *Comstock Women: The Making of a Mining Community*, edited by Ronald M. James and C. Elizabeth Raymond, 246–262. Reno and Las Vegas: University of Nevada Press, 1998.

———. "Knockers, Knackers, and Ghosts: Immigrant Folklore in the Western Mines." *Western Folklore* 51, no. 2 (April 1992): 153–157.

———. *The Roar and the Silence.* Reno and Las Vegas: University of Nevada Press, 1998.

James, Ronald M., and C. Elizabeth Raymond, eds. *Comstock Women: The Making of a Mining Community.* Reno and Las Vegas: University of Nevada Press, 1998.

James, Susan A. "Queen of Tarts." *Nevada Magazine* 44, no. 5 (September–October 1984): 51–54.

Johnson, Susan Lee. *Roaring Camp: The Social World of the California Gold Rush.* New York: Norton, 2000.

Jones, Martin. *The Molecule Hunt: Archaeology and the Search for Ancient DNA.* New York: Arcade, 2001.

Kallet, Arthur, and F. J. Schlink. *100,000,000 Guinea Pigs: Dangers in Everyday Foods, Drugs, and Cosmetics.* New York: Vanguard, 1933.

Kartdatzke, Tim A. *Archaeological Excavations in Skagway.* Vol. 9, *Excavations at the Pantheon Saloon Complex.* Anchorage: National Park Service, 2002.

Kinchloe, Jessica Leigh. "'The Best the Market Affords': Food Consumption at the

Merchant's Exchange Hotel, Aurora, Nevada." Master's thesis, University of Nevada, Reno, 2001.

Knauft, Bruce A. *Genealogies for the Present in Cultural Anthropology.* New York: Routledge, 1996.

Koch, Augustus. "Bird's-eye View of Virginia City, Nevada." San Francisco: Britton, Rey, and Company, 1875.

Kovel, Ralph, and Terry Kovel. *Kovels' New Dictionary of Marks, Pottery and Porcelain: 1850 to the Present.* New York: Crown, 1986.

Lang, William L. "Helena, Montana's Black Community, 1900–1912." In *African Americans on the Western Frontier,* edited by Monroe Lee Billington and Roger D. Hardaway, 198–216. Niwot: University Press of Colorado, 1998.

Leavitt, Robert. "Appendix B. Stoneware 'Selters' Bottle Analysis." In "The Archaeology of Piper's Old Corner Bar, Virginia City, Nevada," by Kelly J. Dixon, with contributions by Ronald M. James, Robert C. Leavitt, Dan Urriola, and Chris Urriola. Comstock Archaeology Center Preliminary Report of Investigations. Carson City: Nevada State Historic Preservation Office, 1999.

———. "Gin Jugs and Mineralwasser: A Taste of Home." Paper presented at the 35th annual meeting of the Society for Historical Archaeology, Mobile, AL, 2002.

———. "Stoneware 'Nassau' Bottles from Germany." Paper presented to the Nevada Archaeological Association, Reno, Nevada, 1999.

———. "Taking the Waters: Stoneware Jugs and the Taste of Home They Contained." Master's thesis, University of Nevada, 2004.

Le Roy Ladurie, Emmanuel. *Montaillou: The Promised Land of Error.* New York: Vintage Books, a Division of Random House, 1979.

Limerick, Patricia Nelson. *The Legacy of Conquest: The Unbroken Past of the American West.* New York: Norton, 1987.

Lord, Eliot. *Comstock Mining and Miners.* 1883. Reprint, Washington, DC: U.S. Geological Survey, Government Printing Office, 1959.

Loverin, Jan I., and Robert A. Nylen. "Creating a Fashionable Society: Comstock Needleworkers From 1860 to 1880." *In Comstock Women: The Making of a Mining Community,* edited by Ronald M. James and C. Elizabeth Raymond, 115–141. Reno and Las Vegas: University of Nevada Press, 1998.

Loy, Thomas. "Prehistoric Blood Residues: Detection on Tool Surfaces and Identification of Species of Origin." *Science* 220 (1983): 1269–1271.

Loy, Thomas, and E. James Dixon. "Blood Residues on Fluted Points From Eastern Alaska." *American Antiquity* 63, no. 1 (1998): 21–46.

Loy, Thomas, and B. L. Hardy. "Blood Residue Analysis of 90,000-Year-Old Stone Tools from Tabun Cave, Israel." *Antiquity* 66 (1992): 24–35.

Loy, Thomas, and A. R. Wood. "Blood Residue Analysis of Cayonu Tepesi, Turkey." *Journal of Field Archaeology* 16 (1989): 451–460.

Mathews, Mary McNair. *Ten Years in Nevada: or, Life on the Pacific Coast.* Lincoln: University of Nebraska Press, 1985.

Matthews, Glenna. "Forging a Cosmopolitan Civic Culture: The Regional Identity of San Francisco and Northern California." In *Many Wests: Place, Culture, and Regional Identity,* edited by David M. Wrobel and Michael C. Steiner, 211–234. Lawrence: University of Kansas Press, 1997.

McNally, Norah, Robert C. Shaler, Alan Giusti, Michael Baird, Ivan Balazs, Peter De Forest, and Lawrence Kobilinsky. "Evaluation of Deoxyribonucleic Acid (DNA) Isolated from Human Bloodstains Exposed to Ultraviolet Light, Heat, Humidity, and Soil Contamination." *Journal of Forensic Sciences* 34, no. 5 (1989): 1059–1069.

Merrifield, Ralph. *The Archaeology of Ritual and Magic.* New York: New Amsterdam Books, 1987.

Miller, George L. "A Revised Set of CC Index Values for Classification and Economic Scaling of English Ceramics from 1787 to 1880." *Historical Archaeology* 25, no. 1 (1991): 1–25.

Mrozowski, Stephen A., Grace H. Ziesling, and Mary C. Beaudry. *Living on the Boott: Historical Archaeology at the Boott Mills Boardinghouses, Lowell, Massachusetts.* Amherst: University of Massachusetts Press, 1996.

Mullins, Paul R. *Race and Affluence: An Archaeology of African America and Consumer Culture.* New York: Kluwer Academic/Plenum Publishers, 1999.

Murphy, Mary Martin. *Mining Cultures: Men, Women, and Leisure in Butte, 1914–1941.* Urbana: University of Illinois Press, 1997.

Nowicki, Jack. "Analysis of Chemical Protection Sprays by Gas Chromatography/Mass Spectroscopy." *Journal of Forensic Sciences* 27 no. 3 (1982): 704–709.

Orser, Charles E., Jr. *A Historical Archaeology of the Modern World.* New York: Plenum, 1996.

———. ed. *Race and the Archaeology of Identity.* Salt Lake City: University of Utah Press, 2001.

Orser, Charles E., Jr., and David W. Babson. "Tabasco Brand Pepper Sauce Bottles from Avery Island, Louisiana." *Historical Archaeology* 25 no. 2 (1990): 107–114.

Orser, Charles E., Jr., and Brian M. Fagan. *Historical Archaeology.* New York: Harper Collins College Publishers, 1995.

Osborne, Peggy Ann. *About Buttons: A Collector's Guide 150 A.D. to the Present.* Lancaster, PA: Schiffer, 1993.

Pfeiffer, Michael A. "The Tobacco-Related Artifact Assemblage from the Martinez

Adobe, Pinole, California." In *The Archaeology of the Clay Tobacco Pipe.* Part 7, *America,* edited by Peter Davey, 185–194. Oxford, England: B.A.R. Series 175, 1983.

Phillips, Thomas D. "The Black Regulars." In *The West of the American People,* edited by Allan G. Bogue, Thomas D. Phillips, and James E. Wright, 138–143. Itasca, IL: Peacock, 1970.

Porter, Kenneth W. "Black Cowboys in the American West, 1866–1900." In *African Americans on the Western Frontier,* edited by Monroe Lee Billington and Roger D. Hardaway, 110–127. Niwot: University Press of Colorado, 1998.

Powdermaker, Hortense. *Hollywood, The Dream Factory: An Anthropologist Looks at the Movie-Makers.* Boston: Little, Brown, 1950.

Praetzellis, Adrian, and Mary Praetzellis. "Mangling Symbols of Gentility in the Wild West." *American Anthropologist* 103, no. 3 (2001): 645–654.

———. *"We Were There, Too": Archaeology of an African-American Family in Sacramento, California.* Cultural Resources Facility, Anthropological Studies Center. Rohnert Park, CA: Sonoma State University, 1992.

———. eds. "Archaeologists As Storytellers." *Historical Archaeology* 32, no. 1 (1998).

Protz, Roger. *Ultimate Encyclopedia of Beer: The Complete Guide to the World's Greatest Brews.* London: Carlton Books, 1995.

Purser, Margaret. "Gender Archaeology." In *Archaeological Method and Theory: An Encyclopedia,* edited by Linda Ellis, 233–237. New York: Garland, 2000.

Rapp, George "Rip," Jr., and Christopher L. Hill, eds. *Geoarchaeology: The Earth-Science Approach to Archaeological Interpretation.* New Haven, CT: Yale University Press, 1998.

Rockman, Diana Diz, and Nan A. Rothschild. "City Tavern, Country Tavern: An Analysis of Four Colonial Sites." *Historical Archaeology* 18 (1984) 2: 112–121.

Rothschild, Nan A. *New York City Neighborhoods: The Eighteenth Century.* New York: Academic Press, 1990.

Rusco, Elmer. *"Good Times Coming?" Black Nevadans in the Nineteenth Century.* Westport, CT: Greenwood, 1975.

Schablitsky, Julie M. "The Magic Wand: Hypodermic Drug Injection of the Nineteenth Century." Paper presented at the 35th annual meeting of the Society for Historical Archaeology, Mobile, AL, 2002.

———. "The Other Side of the Tracks: The Archaeology and History of a Virginia City, Nevada, Neighborhood." Ph.D. diss., Portland State University, 2002.

Schamberger, Hugh A. *Historic Mining Camps of Nevada, Water Supply for the Comstock: Early History, Development, Water Supply.* Prepared in cooperation with Nevada Department of Conservation and Natural Resources and U.S. Geological Survey, 1969.

Schubert, Frank N. "Black Soldiers on the White Frontier: Some Factors Influencing Race Relations." *Phylon* 32 (Winter 1971): 410–415.

Schulz, Peter D., and Sherri M. Gust. "Faunal Remains and Social Status in Nineteenth-Century Sacramento." *Historical Archaeology* 17, no. 1 (1983): 44–53.

Schulz, Peter D., Betty J. Rivers, Mark M. Hales, Charles A. Litzinger, and Elizabeth A. McKee. *The Bottles of Old Sacramento: A Study of Nineteenth-Century Glass and Ceramic Retail Containers.* Part 1. California Archaeological Reports, no. 20. Sacramento: California State of California Department of Parks and Recreation, Cultural Resources Management Unit, 1980.

Scott, Elizabeth M. "Faunal Remains From the Hibernia Saloon and Faunal Remains From the O'Brien and Costello Bar and Shooting Gallery." In *Public Archaeology on the Comstock,* by Donald L. Hardesty, with Jane E. Baxter, Ronald M. James, Ralph B. Giles Jr., and Elizabeth M. Scott, 64–88. University of Nevada, Reno report prepared for the Nevada State Historic Preservation Office. Carson City: Nevada State Historic Preservation Office, 1996.

Shange, Ntozake. *If I Can Cook/You Know God Can.* Boston: Beacon, 1998.

Smith, Duane A. "Comstock Miseries: Medicine and Mining in the 1860s." *Nevada Historical Society Quarterly* 36, no. 1 (Spring 1993): 1–12.

Smith, Raymond M. *Saloons of Old and New Nevada: Commentaries on the Role and Development of the Nevada Saloon.* Minden, NV: Silver State Print, 1992.

Sprague, Roderick. "A Functional Classification for Artifacts From Nineteenth- and Twentieth-Century Sites." *North American Archaeologist* 2, no. 3 (1980): 259.

Stein, Gil J. *Rethinking World-Systems: Diasporas, Colonies, and Interaction in Uruk Mesopotamia.* Tucson: University of Arizona Press, 1999.

Steinberg, Stephen. *The Ethnic Myth: Race, Ethnicity, and Class in America.* Boston: Beacon, 2001.

Swann, June. "Shoes Concealed in Buildings," *Costume Society Journal* 30 (1996): 65–66.

Swerdlow, Joel L. "Changing America." *National Geographic* 200, no. 3 (September 2001): 42–61.

Switzer, Ronald R. *The Bertrand Bottles: A Study of Nineteenth-Century Glass and Ceramic Containers.* Washington, DC: National Park Service, U.S. Department of the Interior, 1974.

Taylor, Quintard. *In Search of the Racial Frontier: African Americans in the American West, 1528–1990.* New York: Norton, 1998.

Twain, Mark [Samuel Clemens, pseud.]. *Mark Twain in Virginia City, Nevada.* Las Vegas: Nevada Publications, 1985.

———. *Roughing It.* 1873. Reprint, New York: Penguin, 1981.

Urriola, Dan L. "Appendix D: Mended Ceramic Artifact Assemblage." In "The Archaeology of Piper's Old Corner Bar, Virginia City, Nevada," by Kelly J. Dixon, with contributions by Ronald M. James, Robert C. Leavitt, Dan Urriola, and Chris Urriola. Comstock Archaeology Center Preliminary Report of Investigations. Carson City: Nevada State Historic Preservation Office, 2001.

Wake, Thomas A. "Appendix K: Faunal Report, Zooarchaeology of the Pantheon Saloon and Its Local Area, Skagway, Alaska." In *Archaeological Excavations in Skagway.* Vol. 9, *Excavations at the Pantheon Saloon Complex,* by Tim A. Kartdatzke, K-1–K-108. Anchorage: National Park Service, 2002.

Waldorf, John Taylor. *A Kid on the Comstock: Reminiscences of a Virginia City Childhood.* Reno: University of Nevada Press, 1970.

Waldstreicher, David. *In the Midst of Perpetual Fetes: The Making of American Nationalism, 1776–1820.* Chapel Hill: University of North Carolina Press, 1997.

Wall, Diana diZerega. *The Archaeology of Gender: Separating the Spheres in Urban America.* New York: Plenum, 1994.

Ward, William Hayes. *The Seal Cylinders of Western Asia.* Washington, DC: Carnegie Institution, 1910.

West, Elliott. *The Saloon on the Rocky Mountain Mining Frontier.* Lincoln: University of Nebraska Press, 1979.

———. *The Way to the West: Essays on the Central Plains.* Albuquerque: University of New Mexico Press, 1995.

Wilson, Rex. *Bottles on the Western Frontier.* Tucson: University of Arizona Press, 1981.

Wiseman, D. J. *Cylinder Seals of Western Asia.* London: Batchworth, 1958.

Wood, Margaret C., Richard F. Carrillo, Terri McBride, Donna L. Bryant, and William J. Convery III. *Historical Archaeological Testing and Data Recovery for the Broadway Viaduct Replacement Project, Downtown Denver, Colorado: Mitigation of Site 5DV5997.* SWCA Archaeological Report No. 99–308. Submitted to the Colorado Department of Transportation, Office of Environmental Services, Denver, Colorado and Hamon Contractors, Inc., Denver Colorado. Westminster, CO; SWCA, Inc., Environmental Consultants, 1999.

Woods, Randall B. "Integration, Exclusion, or Segregation? The 'Color Line' in Kansas, 1878–1900." In *African Americans on the Western Frontier,* edited by Monroe Lee Billington and Roger D. Hardaway, 128–146. Niwot: University Press of Colorado, 1998.

Wright, William [Dan DeQuille, pseud.]. *The Big Bonanza.* 1876. Reprint, New York: Knopf, 1953.

HISTORICAL RECORDS

Directories

Bishop's Directory of Virginia City, Gold Hill, Silver City, Carson, and Reno, 1878–1879. San Francisco: B. C. Vandall.

Collins, Charles. *Charles Collins Mercantile Guide and Directory for Virginia City and Gold Hill, 1864–1865.* Virginia City: Agnew and Deffebach, 1865.

Kelly, J. Wells. *J. Wells Kelly's Second Directory of Nevada Territory, 1863–1864.* Virginia City: Valentine and Company, 1863.

McKenney's Pacific Coast Business Directory 1882. San Francisco: L. M. McKenney and Co.

Pacific Coast Business Directory, 1871–1873. San Francisco: Henry G. Langly, Langley Publishing.

Uhlhurn, John F. *Virginia and Truckee Railroad Directory, 1873–1874.* Virginia City, NV.

Manuscripts

Nevada State Census on Microfilm 1863, 1875

U.S. Manuscript Census on Microfilm 1860, 1870, 1880

Newspapers

Carson Daily Appeal

Daily Stage

Footlight

Gold Hill Daily News

Montana Plaindealer

Pacific Appeal

Territorial Enterprise

Virginia Evening Bulletin

Virginia Evening Chronicle

INDEX

Note: Italic page numbers refer to illustrations.